In Vitro Methods for Conservation of Plant Genetic Resources

In Vitro Methods for Conservation of Plant Genetic Resources

Edited by

JOHN H. DODDS

Head of Genetic Resources Department
International Potato Centre
Lima, Peru

CHAPMAN AND HALL

LONDON · NEW YORK · TOKYO · MELBOURNE · MADRAS

UK   Chapman and Hall, 2–6 Boundary Row, London SE1 8HN

USA   Van Nostrand Reinhold, 115 5th Avenue, New York NY10003

JAPAN   Chapman and Hall Japan, Thomson Publishing Japan, Hirakawacho
            Nemoto Building, 7F, 1–7–11 Hirakawa-cho, Chiyoda-ku, Tokyo 102

AUSTRALIA   Chapman and Hall Australia, Thomas Nelson Australia, 102 Dodds Street,
            South Melbourne, Victoria 3205

INDIA   Chapman and Hall India, R. Seshadri, 32 Second Main Road, CIT East,
            Madras 600 035

First edition 1991

© 1991 Chapman and Hall

Typeset in 10/12 Meridien by
Rowland Phototypesetting Ltd, Bury St Edmunds
Printed in Great Britain at
The University Press, Cambridge

ISBN 0 412 33870 X (HB)
ISBN 0 442 31165 6 (USA)

British Library Cataloguing in Publication Data

In vitro methods for conservation of plant genetic
resources.
1. Plants. Genetics
I. Title
581.15

ISBN 0-412-33870-X

**Library of Congress Cataloging-in-Publication Data
Available**

Contents

Contributors

M. BRAVATO
Instituto Internacional de Estudios Avanzados (IDEA),
Unidad de Biotecnologia,
Caracas,
Venezuela

J. H. DODDS
International Potato Center (CIP),
Lima,
Peru

Z. HUAMAN
International Potato Center (CIP),
Lima,
Peru

M. G. K. JONES
Rothamsted Experimental Station,
Department of Biochemistry,
Harpenden, Herts,
UK

C. G. KUO
Asian Vegetable Research and Development Center (AVRDC),
Tainan, Taiwan,
China

R. LIZARRAGA
International Potato Center (CIP),
Lima,
Peru

N. Q. NG
International Institute of Tropical Agriculture (IITA),
Ibadan,
Nigeria

Contributors

S. Y. C. NG
International Institute of Tropical Agriculture (IITA),
Ibadan,
Nigeria

R. H. POTTER
Rothamsted Experimental Station,
Department of Biochemistry,
Harpenden, Herts,
UK

L. E. TOWILL
USDA/ARS,
National Seed Storage Laboratory,
Fort Collins, Colorado,
USA

L. VILLEGAS
Instituto Internacional de Estudios Avanzados (IDEA),
Centro de Educacion,
Caracas,
Venezuela

C. P. WILKINS
Department of Cell and Structural Biology
University of Manchester,
Manchester,
UK

1

Introduction: Conservation of plant genetic resources – the need for tissue culture

J. H. DODDS

1.1 HISTORY AND PRINCIPLES OF PLANT GENETIC CONSERVATION

In the 1920s, N. I. Vavilov and his collaborators at the USSR Institute of Plant Industry located and described what are today known as the geographical centres of genetic diversity of our cultivated plants and their wild relatives.

These early workers discovered a vast wealth of previously unknown genetic variation, providing valuable resources for use in genetic, cytogenetic and evolutionary studies and giving to plant breeders vast reserves of new and original material. However, although early warnings of the gradual disappearance of some primitive crop varieties in such 'gene centres' were made as early as 1936, little attention was paid to the necessity to preserve these reservoirs of genetic diversity or 'natural gene pools'.

Similarly, the recommendations of the first FAO Technical Conference on Plant Exploration and Introduction held in 1961 also failed to convey a sense of urgency. It was not until the mid-1960s, following urgent warnings of a greatly accelerated rate of displacement of primitive crop varieties by new, introduced cultivars, that definite action was at last taken. These primitive crop races had evolved into complex diverse populations through long exposure to environmental stresses and cultivation over long periods of time, and by competing and introgressing with associated crop, wild and weed species. These complex gene blocks were to provide the basic material used to build our modern high-yielding cultivars.

In Vitro *Methods for Conservation of Plant Genetic Resources. Edited by John H. Dodds.* *Published in 1991 by Chapman and Hall, London.* ISBN 0 412 33870 X

However, the transition from primitive to 'advanced' cultivars has had the effect of narrowing the genetic base. This has happened in two distinct ways: (1) selection for relative uniformity, resulting in 'pure' lines, multi-lines, single or double hybrids, etc.; and (2) selection for closely defined objectives. Both of these processes have resulted in a marked reduction in genetic variation. At the same time, there has been a tendency to restrict the gene pool from which parental material has been drawn. This is a result of the high level of productivity achieved when breeding within a restricted but well-adapted gene pool, and of breeding methods which have made it possible to introduce specifically desired improvements, such as disease resistance and quality characteristics, into breeding stocks with a minimum of disturbance to genotypic structure.

Developments in agriculture, such as intensive mechanization, the widespread application of fertilizers and the use of herbicides, fungicides and pesticides, have created a situation whereby a few, selected high-yielding cultivars may be grown over large parts of the earth, so further contributing to a decline in crop genetic diversity. This process is under way in all countries, both developed and developing, and unfortunately includes some of the richest primary and secondary gene centres of several important food crops. This process of gradual erosion can be exemplified by considering two examples: (1) the highly successful wheats produced by the Rockefeller team in Mexico which transformed the pattern of agriculture over much of Asia and Latin America; and (2) the succession of new rice varieties produced by the International Rice Research Institute (IRRI) in the Philippines.

In 1964, the International Biological Programme (IBP) of the International Council of Scientific Unions (ICSU) was initiated, and a subcommittee was set up to study ways and means of collecting and conserving plant genetic resources which were threatened by agricultural developments in many of the centres of genetic diversity. The following year there occurred an integration of effort between the IBP and the Food and Agriculture Organization of the United Nations (FAO) into the problem of conservation of plant genetic resources. This collaboration resulted in a joint FAO/IBP Technical Conference being held at FAO headquarters in Rome during September 1967. The aim of this enquiry was to examine the current availability of plant genetic resources, and the status and organization of their exploration, evaluation, utilization and conservation. The proceedings of this conference were published as part of a series of IBP handbooks. The conference listed five distinct kinds of germplasm material which are in need of preservation: (1) cultivars in current use; (2) obsolete cultivars; (3) special genetic stocks such as resistance stocks, genetic and cytogenetic material, and induced mutations; (4) primitive varieties or land races; and (5) wild and weed species related to cultivated species.

Detailed proposals were discussed for the conservation of genetic resources both as natural mass reservoirs in which intact wild communities could be preserved, such as the Hills of Eastern Galilee in Israel, and as 'crop reservations', where local primitive cultivars or land races could be preserved in their original habitats. However, there are obvious arguments, both technical, political and financial, against this type of conservation. Another proposal called for the setting up of specific collections of genotypes (gene banks) representative of a region or of the world. Specific methods of conservation as applied to different crop species were discussed, and the use of seed-storage banks was proposed as a means of conservation of certain crops. Also, for the first time, the difficulties of conserving long-lived, large organisms were highlighted; and prospects were explored for the possible conservation of such resources by tissue-culture techniques. However, emphasis was placed on the need for a great deal of research and development before such ideas could become reality.

The 1967 conference did much to define the goals and strategies of plant genetic conservation, and to develop the required scientific background and methodology. A survey of genetic resources in the field was to supply urgently required information on priorities for exploration. Long-term conservation of seeds, under appropriate conditions, was recommended, where applicable, as the safest and most economical method of conservation, and a co-operative network of seed storage laboratories was seen as the most efficient and economical form of organization.

More recently, the FAO Expert Panel on Plant Exploration and Introduction met in 1969, 1970 and 1973; its reports further elaborated the major recommendations of the 1967 conference. Proposals were drafted for an international network of genetic resources centres, and the function and responsibilities of participating institutions were defined.

Also at around this time (1972) the United Nations Environment Conference at Stockholm served to highlight the problems of the world's genetic resources, and the need to save and preserve irreplaceable genetic resources for the good of future generations. This was followed by a decision of UNEP, the United Nations Environment Programme, to provide financial support for the exploration and conservation of genetic resources. Subsequently, the Consultative Group on International Agricultural Research (CGIAR), after two years of study and negotiation, adopted the proposal of its Technical Advisory Committee to support a global network of genetic resources centres, with special emphasis being given to support for centres in the regions of genetic diversity. Funds for exploration, conservation, documentation and training were allocated from a central fund on a project basis, and an International Board for Plant Genetic Resources (IBPGR) was set up to administer the fund.

At the 1974 FAO/IBP Technical Conference, many aspects of plant

3

genetic conservation were discussed, including: exploration, evaluation, conservation and storage, documentation/information management and technical aspects of genetic resources centres. A great deal of attention was paid to methods of conservation and storage; and several solutions to problems of conservation previously discussed at the 1967 Technical Conference were proposed.

In particular, a comprehensive review by Roberts (1975) of the principles and methods of seed and pollen storage demonstrated that long-term seed storage was a relatively simple and inexpensive operation in terms of technology, facilities, staff and operating expenses. It was evident that seeds of most crop plants could be stored over long periods of time with a minimum risk of genetic damage. Additional reports by Villiers (1975) and Sakai and Nishero (1975) further indicated that seeds of certain crop species which could not be stored via conventional techniques, i.e. low temperature ($-20°C$) and low moisture content (5%), and which were described as 'recalcitrant' by Roberts, could possibly be stored either in a fully imbibed state, or at the temperature of liquid nitrogen ($-196°C$). In addition, work was presented by Morel (1975) and Henshaw (1975) which illustrated the potential for applying tissue-culture techniques to the problems of storage of certain categories of germplasm resources, and emphasized the enormous advances in tissue-culture technology which had been made in the six years since the previous IBP/FAO Technical Conference.

Since the 1974 conference, many National and International seed banks and genetic resources centres have been established, and some are still under construction. Recently, several comprehensive IBPGR Technical Reports have been published, covering many aspects of seed conservation, including: (1) conventional seed-storage techniques; (2) storage of recalcitrant seeds; and (3) technical aspects of gene-bank construction and management.

1.2 THE APPLICATION OF TISSUE-CULTURE TECHNIQUES TO STORAGE OF GENETIC RESOURCES AND MICROPROPAGATION

A great deal of interest has been expressed in recent years concerning the application of tissue-culture or *in vitro* techniques to plant genetic conservation. Some of the reasons for this upsurge of interest have already been mentioned, i.e. the inherent 'recalcitrance' of seeds of many important crops, including many timber, food and commercial crops such as *Elaeis guineensis* (oil palm), *Hevea brasiliensis* (rubber), *Citrus* spp. and *Coffea* spp.

Other reasons why tissue-culture techniques could be of great value for germplasm conservation are discussed below.

Vegetative propagated plants which have a high degree of heterozygosity, or which do not produce seed, must be stored in a vegetative form as tubers, roots, cuttings, etc. This usually involves a costly (in financial and labour terms) annual cycle of growth and field collection and/or storage of propagules for short periods. Both of these methods are prone to possible catastrophic losses due to: (1) attack by pests and pathogens; (2) climate disorders; (3) natural disasters; and (4) political/economic causes. Crops which fall into this category include yams (*Dioscorea* spp.), bananas (*Musa* spp.), potatoes (*Solanum* spp.), cassava (*Manihot* spp.), taro (*Colacasi esculentum*) and sweet potato (*Ipomoea batatas*).

For many palm species, including coconut (*Cocos nucifera*), oil palm (*Elaeis guineensis*) and date palm (*Phoenix dactylifera*), there is no efficient method of vegetative propagation. For example, the conventional method of reproduction of date palm, i.e. via suckers arising from the base of the main stem, is both slow and unreliable. The development and application of a suitable tissue-culture system of propagation and conservation would therefore have obvious benefits. Again, in the case of date palm the situation is especially serious, as date palm genomes in Algeria and Morocco are bordering on extinction due to Bayoud disease (*Fusarium oxysporum* Schlect. var. *albedinis*).

Many long-lived forest trees, including both Angiosperms and Gymnosperms, do not produce seed until a certain age, and must be propagated vegetatively when it is desired to reproduce the parental genotype. With Rosaceous fruit trees, the most frequently used methods of vegetative propagation at present are budding and grafting. However, one area of research currently being investigated is the possibility of very rapid clonal propagation of fruit trees by means of tissue-culture techniques. These techniques have special relevance to certain fruit-tree cultivars which are either difficult and/or expensive to propagate by conventional means. For example, apple scion cultivars are usually propagated by budding or grafting onto rootstocks, which are themselves raised by stooling or layering. This process takes three years and demands expensive nursery facilities and skills.

Propagation by the production of self-rooted plants would be much more rapid, and could be of value in hastening the availability of new cultivars from breeding systems. An additional application of tissue-culture techniques to fruit-tree propagation is the rapid *in vitro* multiplication and rooting of apple rootstocks such as M.9 (see Chatper 8). This rootstock is widely used because of its effects on precocity and the control of tree and fruit size. A major disadvantage, however, is the difficulty of rooting cuttings by conventional methods.

1.3 THE *IN VITRO* CULTURE OF PLANTS

The potential for regenerating a whole plant from a single cell was first hypothesized by the German botanist Haberlandt in 1902; however, it was not until the 1950s that whole plants were regenerated from isolated cultured cells. In the 30 years or so since Haberlandt's principle of the 'totipotency' of single plant cells was realized, the field of plant tissue, cell and organ culture has grown to be of immense size, importance and scope.

It would be pointless to try to review or even pretend to give an introduction to the many facets of plant tissue culture, and it will suffice merely to point out that several textbooks are available which fully cover all aspects of the basic principles and methodology of plant tissue culture (Thomas and Davey, 1975; Dodds and Roberts, 1984). In addition, a number of tissue-culture periodicals now exist, and specialized texts are available which provide comprehensive discussions covering all aspects of the application of plant tissue culture to such diverse topics as plant improvement, plant propagation, somatic hybridization, genetic engineering, biosynthesis of secondary products and plant pathology.

The term *in vitro* or tissue culture covers a wide range of techniques involving the growth, under sterile conditions, of whole plant organs such as shoots or embryos, or alternatively the culture of masses of unorganized callus or single cells, or even cells devoid of cell walls, known as protoplasts. Much interest has recently been shown in the use of such 'naked cells' in studies concerned with genetic engineering and somatic hybridization. Techniques of tissue culture can be broadly categorized as follows: (1) organ culture; (2) callus culture; (3) suspension cultures; (4) isolation and culture of single cells and protoplasts; and (5) anther and pollen culture.

In applying each of these techniques to the special problems posed by plant genetic conservation, it is evident that some of the techniques are more readily applicable than others, and certain of the categories are associated with special difficulties which have yet to be overcome. The ultimate aim of any tissue-culture system employed for the purposes of genetic conservation is that of the preservation of specific and unique individual genomes. One of the major advantages of a perfect tissue-culture system involving the culture of either callus or a cell suspension is that, in a model system, one can culture a small piece of callus, or volume of cell suspension, to produce any desired quantity of cells, and then merely change the culture conditions to induce either somatic embryogenesis or adventitious shoot formation and so obtain the desired number of regenerated plants. Ideally, therefore, one would wish to conserve a particular genome in the form of single cells, or a piece of callus; then when that specific genome was required – e.g. by a plant breeder – it would be a simple matter to regenerate the desired quantity of plants. In practice, though, except for a few model systems which have little useful application, such an idealized approach is not attainable for most plants at the present time. However, past experience indicates that it may simply be a matter of time before such perfect culture systems can be elucidated and applied to a wide range of important crop plants.

There are also other factors which discourage the use of unorganized tissue cultures for purposes of conservation; these are the phenomenon of genetic instability of some culture systems, and the observation that, with certain tissue cultures, the morphogenetic potential of the cultures may decline after growth under *in vitro* conditions for an extended period of time. A report by Chandler and Dodds (1982) has indicated that the decline in morphogenetic potential observed in callus cultures of *Solanum laciniatum* with increasing time in culture can be correlated with an increase in the levels of ploidy within the cells. Genetically unstable cultures may also arise as a result of initiating cultures from plants whose tissues are misoploid. Some of the details and implications of the genetic instability of plant tissue cultures have been reviewed by D'Amato (1980), whilst some of the difficulties encountered in the use of cell and callus cultures for genetic conservation have been discussed by Withers (1980) and Henshaw *et al.* (1979). Both of these authors concluded that cultures derived from apical-meristem tissues should be the preferred choice of starting material both for the production of clones and as specimens for germplasm storage.

In spite of the limitations discussed above, the benefits to be gained from the use of tissue-culture systems can be enormous, and are listed below:

1. Most *in vitro* systems possess the potential for very high multiplication rates; for example, Morel (1975) states that with both orchids and grapes it is possible to obtain several million plants in the space of one year from a single meristem explant. Similarly Jones *et al.* (1979) reported that, by means of shoot-tip culture of apple, it was possible to obtain several tens of thousands of clonal plants per year.
2. Tissue-culture systems are aseptic, and can easily be kept free from fungi, bacteria, viruses and insect parasites. One important use of tissue culture at present is the culturing of meristems so as to obtain virus-free plants, and the subsequent storage of such plants as phathogen-free stocks by plant breeders.
3. Space considerations: Galzy (1969) demonstrated that, by storing plantlets of grape (*Vitis rupestris*) at a low temperature (9°C) for periods of one year, a total of 800 cultivars with six replicates per cultivar could be stored in an area of 2 m². The comparable *in vivo* situation would have required 1 ha of field space.
4. In an ideal tissue-culture storage system, genetic erosion is reduced to zero.

5. By means of specialized *in vitro* techniques, such as pollen and anther culture, haploid plants may be produced which are of use in breeding programmes.
6. Tissue-culture conservation techniques are also useful in plant-breeding programmes as a means of rescuing and subsequently culturing zygotic embryos from incompatible crosses which normally result in embryo abscission. Such a technique has been successfully applied to avocado.
7. The expense, both in labour and financial terms, of maintaining large field collections is a further factor contributing to the use of *in vitro* collections.

This text will highlight both the techniques and the major crops which have made tissue culture a useful tool for the genetic resources specialist.

REFERENCES

Chandler, S. F. and Dodds, J. H. (1982) Solasodive production in rapidly proliferating tissue cultures of *Solonium lacincatum*. *Plant Cell Reports*, **2**, 69–72.

D'Amato, F. (1978) in *Frontiers of Plant Tissue Culture* (ed. T. A. Thorpe), International Association of Plant Tissue Culture, Calgary, pp. 89–107.

D'Amato, F. (1980) in *Plant Cell Cultures: Results and Perspectives* (ed. F. Sala, B. Parisi, R. Cella and O. Cifferri), Elsevier, Holland, pp. 287–96.

Dodds, J. H. and Roberts, L. W. (1984) *Experiments in Plant Tissue Culture*, 2nd edn, Cambridge University Press, Cambridge, UK.

Galzy, R. (1969) Recherchez sur la croissance de vitis rupestris scheele sain et court noice cultivé *in vitro* à differentes temperatures. *Am. Phytopathol.*, **1**, 149.

Henshaw, G. G. (1975) in *Crop Genetic Resources for Today and Tomorrow* (ed. O. H. Frankel and J. G. Hawkes), Cambridge University Press, Cambridge, UK, pp. 349–58.

Henshaw, G. G., Stamp, J. A. and Westcott, R. J. (1979) in *Proceedings Plant Tissue Culture Conference* (ed. F. Sala and R. Cella), Blackwell Scientific, Oxford, pp. 186–92.

Jones, O. P., Pontikis, G. A. and Hopwood, M. E. (1979) *J. Hort. Sci.*, **54**, 155.

Morel, G. (1975) in *Crop Genetic Resources for Today and Tomorrow* (ed. O. H. Frankel and J. G. Hawkes), Cambridge University Press, Cambridge, UK, pp. 327–46.

Roberts, E. H. (1975) in *Crop Genetic Resources for Today and Tomorrow* (ed. O. H. Frankel and J. G. Hawkes), Cambridge University Press, Cambridge, UK, pp. 317–26.

References

Sakai, A. and Nisharo, M. (1975) in *Crop Genetic Resources for Today and Tomorrow* (ed. O. H. Frankel and J. G. Hawkes), Cambridge University Press, Cambridge, UK, pp. 317–26.

Street, H. E. (1979) *Plant Cell and Tissue Culture*, Blackwell Scientific, London.

Thomas, E. and Davey, M. R. (1975) *From Single Cells and Plants*, Wykeham, London.

Villiers, T. A. (1975) in *Crop Genetic Resources for Today and Tomorrow* (ed. O. H. Frankel and J. G. Hawkes), Cambridge University Press, Cambridge, UK, pp. 287–317.

Withers, L. A. (1980) *Tissue Culture Storage for Genetic Conservation*, IBPGR Technical Document, Rome.

2

Reduced-growth storage of germplasm

S.Y.C. NG and N.Q. NG

2.1 INTRODUCTION

It is imperative to maintain and preserve germplasm collections represent-ing a wide array of genetic diversity of cultivated plants and their wild relatives for future plant breeding. The need for preserving such genetic stocks for future use cannot be overemphasized.

The existing germplasm banks for plants are evidenced by seed collec-tions. The vast majority of plant species have seeds that are orthodox. The viability of these seeds can be extended by lowering their moisture content and storing at a low temperature. For most species, the lower the seed moisture content, to a lower limit of 3–5%, and the lower the storage temperature ($-20°C$), the longer the seeds can be stored. Nevertheless, several groups of economically important plants cannot be preserved using such storage techniques. There are some plant species, such as coffee, cocoa, oil palm and coconut, where the seeds are recalcitrant and cannot be stored for long periods of time. In this case, decreases in seed moisture content even at a relatively high level, 12–13%, or exposure to low storage temperatures, tend to decrease the period of viability.

For vegetatively propagated plants, such as the root and tuber crops, germplasm conservation presents special problems. Germplasm of these crops is usually maintained in the form of tubers, roots, bulbs and cuttings, and collections of germplasm are kept in the field as living collections. Maintaining germplasm collections in the field is expensive not only in terms of labour, land and space, but also in terms of loss of valuable genetic material due to disease and pest attack and other unforeseen problems of maintaining germplasm of these problematic plant species. The *in vitro* reduced-growth storage method offers an immediate solution for short- to medium-term storage of germplasm of these vegetatively

In Vitro *Methods for Conservation of Plant Genetic Resources. Edited by John H. Dodds.*
Published in 1991 by Chapman and Hall, London.   ISBN 0 412 33870 X

propagated crops, whereas cryopreservation offers a solution for long-term conservation.

The principle of reduced-growth storage is based on the manipulations of culture conditions/culture media to allow the cultures to remain viable, but with a growth rate that is very slow. The main advantages of this method are that culture deterioration can be detected visually and therefore loss of viability can be avoided; cultivars can be pathogen-tested, so materials will be available for international exchange; the space required for the storage of *in vitro* materials is relatively small as compared to that needed for field cultivation; and the multiplication rate is high and losses of germplasm due to natural disasters can be avoided. Working on *Lotus corniculatus*, Tomes (1979) reported that storage of 100 genotypes can be accomplished using $0.24 \, m^2$ of shelf space, while the equivalent growth room and field space needed for the same number of plants would be $18 \, cm^3$ and $400 \, m^3$, respectively. He also reported that 34 genotypes out of 100 under study were lost in a field-maintenance nursery during one winter, whereas none of the 100 genotypes maintained *in vitro* were lost.

By using reduced-growth storage methods, the International Institute of Tropical Agriculture (IITA) has now maintained over 1500 accessions of its sweet potato, yam and cocoyam germplasm collections (Ng, 1986a). Based on our experience, it is estimated that 3500 accessions each with ten tubes (plantlets) can be accommodated in a culture room of area $3.2 \times 2.4 \times 2.9 \, m^3$.

Reduced-growth storage has been widely used to maintain differentiated cultures (plantlet and/or shoot) but has been applied to a lesser extent to undifferentiated cultures (callus and suspension cultures).

The reduced-growth storage methods can be divided into four types; the reduction of incubation temperature, the manipulation of culture media, the combination of the former two methods, and the modification of the gaseous environment.

This chapter reviews and describes the reduced-growth storage methods used for plant germplasm conservation.

2.2 LITERATURE REVIEW

A previous survey of 32 crop species of economic importance by Withers (1982) indicated that a great deal of work on *in vitro* germplasm conservation has been done by many institutions and universities, but the majority of the work has not been published. The report also indicated that 19 species of the 32 species surveyed have been able to be maintained by reduced-growth storage methods. Of these 19 species, only five were reported in the literature. The present review includes the species that have not been reported by Withers.

Updated information shows that 49 plant species have thus far been successfully conserved by reduced-growth storage methods. Table 2.1 provides information on the species, type of explant used, culture type, storage method and storage duration of these 49 species. Among the four types of reduced-growth storage methods, the reduction of incubation temperature is the most commonly used, followed by manipulation of culture media and the combination of the above two methods. The modification of gaseous environment was only applied to three species.

The types of culture used for germplasm conservation were mainly shoot/plantlets. Only in a very limited number of cases were callus, suspension or root cultures used. Node cuttings and shoot tips were the two most commonly used explant materials for culturing.

From this review, it is evident that the use of reduced-growth storage methods to conserve germplasm, as well as the number of plant species that can be conserved by this method, have increased over the years. This means that there are more and more researchers working on this area. The importance of the reduced-growth storage method for germplasm conservation has also gained wide recognition.

2.3 METHODS

The ability to develop *in vitro* culture media that can regenerate plants from tissues/organs of a particular plant is of paramount importance in the application of reduced-growth storage methods for germplasm conservation. Other determinant factors, such as the selection of suitable plant materials, culture environments or culture types, will also play an important role in the success of applying such techniques to germplasm conservation.

In this section, we shall describe and review the types of plant materials, surface disinfection processes, culture media and culture conditions that should be considered for establishing cultures and/or for conserving germplasm.

2.3.1 Plant materials

The explant materials used for germplasm conservation should be able to maintain the highest degree of genetic stability. Organized cultures (direct shoot or plant regeneration) should be used unless the conservation is aiming at maintaining cell lines where unorganized cultures (callus and suspension culture) are used. One can consider using unorganized cultures for germplasm conservation if the regeneration of plantlets from this type of culture is well developed. However, the use of unorganized cultures

Table 2.1 Summary of the application of reduced-growth storage in maintaining plant tissue culture

Species	Explant	Culture type	Storage method	Storage duration	Reference
A. Reduced incubation temperature:					
Vitis rupestris Scheele	Shoot	Shoot	At 9°C	10 months	Galzy (1969)
Fragaria annanasus	Meristem	Plantlet	At 1°C or 4°C in darkness. In liquid media with the addition of 1–2 drops of media when needed	6 years	Mullin and Schlegel (1976)
Solanum spp.	Node cuttings	Shoot	On MS medium with 3% sucrose stored at 6°C	12–24 months	Westcott *et al.* (1977)
Beta vulgaris	Axillary shoot (5 mm long) *In vitro* leaf (10–15 mm long)	Plantlet	At 12°C	Over 18 weeks	Hussey and Hepher (1978)
Ipomoea batatas	Node	Plantlet	On MS media with 3% sucrose at 22°C	55 weeks	Alan (1979)
Malus domestica Borkh	Shoot tip	Shoot	At 1 or 4°C	12 months	Lundergan and Janick (1979)
Lotus corniculatas	Node/shoot tip	Plantlet	At 2–4°C	1 month	Tomes (1979)
Trifolium spp.	Shoot tip	Plantlet	At 2–6°C with 300 lux light intensity and 8 h photoperiod or at 4–6°C in complete darkness	15–18 months	Cheyne and Dale (1980)
Medicago sativa Luerne	Shoot tip	Plantlet		15–18 months	Cheyne and Dale (1980)
Lolium spp.	Meristem	Plantlet	At 2–4°C with 300 lux light intensity and 8 h photoperiod	12 months	Dale (1980)
Dactylis glomerata	Meristem	Plantlet		12 months	Dale (1980)
Festuca spp.	Meristem	Plantlet		12 months	Dale (1980)
Phleum pratense	Meristem	Plantlet		12 months	Dale (1980)
Xanthosoma brasiliense	Buds or meristem	Plantlet	At 6°C under dark conditions	More than 2 months	Staritsky (1980)

14

Species	Explant	Morphology	Conditions	Duration	Reference
Trifolium repens cv. grassland Huia	Shoot	Shoot	At 5°C under dark conditions	10 months	Bhojwani (1981)
Manihot esculenta	Node	Plantlet	At 20°C	18–24 months	Centro Internacional de Agricultura Tropical (1981)
Beta vulgaris	Shoot	Plantlet	At 5–10°C with low light intensity	12 months	Miedema (1982)
Beta vulgaris	Flower buds	Shoot	At 5°C with low light intensity	13 months	Miedema (1982)
Citrus spp.	—	Callus	At 4 or 5°C	18 weeks	Withers (1982)
Ficus spp.	—	Callus and suspension	At 15°C	—	Withers (1982)
Phoenix dactylifera	—	Callus	At 5°C	12–18 weeks	Withers (1982)
Saccharum spp.	Leaf sheath tissue	—	At 0–4°C	6 months	Withers (1982)
Theobroma cacao	—	Callus	At reduced temperature (unspecified)	—	Withers (1982)
Vitis rupestris	—	Shoot/plantlet	At 4–12°C	24–40 weeks	Withers (1982)
Asparagus officinalis L.	Mesophyll cell	Cell suspension	At 4°C under dark	15–45 days	Jullien (1983)
Atropa belladonna	Leaf	Callus	At 4°C	3–5 months	Hiraoka and Kodama (1984)
Atractylodes lancea	Leaf	Callus	At 4°C	3–5 months	Hiraoka and Kodama (1984)
Datura innoxia	Hypocotyl	Callus	At 4°C	3–5 months	Hiraoka and Kodama (1984)
	Root	Callus	At 4°C	3–5 months	Hiraoka and Kodama (1984)
	Stem	Callus	At 4°C	3–5 months	Hiraoka and Kodama (1984)
Perilla frutescens	Leaf	Callus	At 4°C	3–5 months	Hiraoka and Kodama (1984)

15

Table 2.1 *Continued*

Species	Explant	Culture type	Storage method	Storage duration	Reference
Bupleurum falcatum	Leaf	Callus	At 4°C	3–5 months	Hiraoka and Kodama (1984)
Dioscorea takoro	Leaf	Callus	At 4°C	3–5 months	Hiraoka and Kodama (1984)
Lithospermum erythrohizon	Leaf	Callus	At 4°C	3–5 months	Hiraoka and Kodama (1984)
Phytolacca americana	Leaf		At 4°C	3–5 months	Hiraoka and Kodama (1984)
Mallatous joponicus	Leaf		At 4°C	3–5 months	Hiraoka and Kodama (1984)
Musa spp.	Meristem	Multiple shoot	At 15°C with 1000 lux light intensity	13–17 months	Banerjee and De Langhe (1985)
Prunus spp.	Shoot	Plantlet	At −3°C under dark	At least 10 months	Marino *et al.* (1985)
Dioscorea rotundata	Node	Plantlet	At 18–22°C (day/night)	1–2 years	Ng and Hahn (1985)
D. alata	Node	Plantlet	At 18–22°C (day/night)	1–2 years	Ng and Hahn (1985)
Populus alba × *P. grandidentata*	Node	Shoot	At 4°C under darkness	12 months	Chun and Hall (1986)
Colocasia esculenta	Buds or meristem	Plantlet	At 9°C	3 years	Zandvoort and Staritsky (1986)
Xanthosoma spp.	—	Plantlet	At 13°C	3 years	Zandvoort and Staritsky (1986)

B. Manipulation of culture media

Secale cereals	Root	Root	Media with 30 mg/l yeast extract subculture every 14 days	74 weeks	Roberts and Street (1955)
Daucus carota	Root disc	Callus	Culture maintained on White's media	116 weeks	Reinert and Backs (1968)

16

Species	Explant	Product	Method	Duration	Reference
Ipomoea batatas	Node	Plantlet	On Heller's medium + 3% sucrose, MS or Nitsch's medium with 1% sucrose	89 weeks	Alan (1979)
Coffea arabica L.	Meristem	Plantlet	On half-strength MS medium without sucrose with 1M IBA maintained at 26°C and 7500 lux light intensity 16 h photoperiod	Over 2 years	Kartha et al. (1981)
Carica papaya	Shoot tip	—	Minimal growth medium	10–12 weeks	Withers (1982)
Coffea spp.	Meristem	Plantlet	Sucrose-limited medium	1 year	Withers (1982)
Passiflora caerulea	Shoot tip	—	Minimal medium	5 weeks	Withers (1982)
Vitis spp.	—	Shoot/plantlet	Medium with low sugar or sugar-free	—	Withers (1982)
Xanthosoma sagittifolium	Corm pieces	Plantlet	White's media incubated at 29°C	1 year	Acheampong and Henshaw (1984)
Vanilla planifolia Andrews	Shoot tip	Plantlet	Pregrowth on MS + 0.5 ppm BAP then transfer to MS basal medium without hormone in screw-cap containers	10 months	Jarret and Fernandez (1984)
Rubus idaeus	Root from shoot-tip culture	Root	Liquid medium consisting of Anderson's basal medium and vitamins with 0.5 mg/l IBA. Cultures are incubated at 24°C under dark conditions	4–6 weeks	Borgman and Mudge (1986)
Lycopersicon esculenta	Axillary shoot	Plantlet	On 75% MS salt with 5% sucrose at culture temperature 30/20°C (day/night)	Reduction in plant growth	Schnapp and Preece (1986)

Table 2.1 *Continued*

Species	Explant	Culture type	Storage method	Storage duration	Reference
Dianthus caryophyllus	Axillary shoot	Plantlet	On 25% or 50% MS salt with 5% sucrose at culture temperature 30/20°C (day/night)	Reduction in plant growth	Schnapp and Preece (1986)
Vitis longii	Anther and ovary	Somatic embryo from callus culture	MS medium without growth regulators and via selective isolation of embryo and subculturing callus	Over 20 months with monthly transfer	Gray and Mortensen (1987)
C. Combination of A and B					
Chrysanthemum morifolium	Internode section	Callus	MS medium with 10% sucrose incubated at 27°C, 16 h photoperiod for 10 days then at 4.5°C for 2 weeks and stored at −3.5°C	28 days	Bannier and Steponkus (1972)
Solanum spp.	Nodal cutting	Plantlet	MS + 2% sucrose + 50 mg *N*-dimethyl succinamic acid. Pre-growth at 20–22°C, 16 h photoperiod at 4000 lux light intensity then stored at 10°C, 16 h photoperiod at 2000 lux light intensity	2 years	Mix (1982)
Solanum spp.	Node	Plantlet	On MS medium with 4% mannitol and stored at 8°C	2–3 years	Espinoza *et al.* (1984)

I. batatas	Node	Plantlet	MS with 3% sucrose and 3% mannitol stored at 18–22°C (day/night)	1–2 years	Ng and Nahn (1985)
Beta vulgaris	Meristem	Shoot	Stored at 4–5°C with 10% sucrose in the medium	2 years	Atanassov (1986)
Actinidia chinensis Planch. cv. Hagwood	Shoot tip	Shoot	1/4 MS salt, 2.7% sucrose, 25 mg/l inositol, 0.1 mg/l thiamine hydrochloride, 4 mg/l BAP and 7 g/l agar stored at 8°C in darkness	52 weeks	Monette (1986)
D. Modification of gaseous environment					
Vitis spp.	—	Callus	Mineral oil added to the medium (agar) up to a maximum depth of 33 mm and stored under low light conditions	120 days	Caplin (1959)
Daucus carota	Phloem	Callus		—	Caplin (1959)
Nicotiana tabacum	Callus	Callus	Stored at oxygen partial pressure below 50 mmHg, growth rate decreased	—	Bridgen and Staby (1981)
Nicotiana tabacum	Shoot	Plantlet		—	Bridgen and Staby (1981)
Chrysanthemum morifolium	Shoot	Plantlet	Stored at oxygen partial pressure below 50 mmHg, growth rate decreased	—	Bridgen and Staby (1981)

should not be encouraged, due to the genetic instability of the regenerated plantlets. Among the explant materials, meristem (shoot tip) and axillary buds are considered to give higher assurance of genetic stability. Very often, meristems with one to two leaf primordia are used not only for the benefit of genetic stability but also for the possibility of obtaining disease-free plants, which can then be used for international exchange. The reasons why virus infections can be eliminated by meristem culture are not yet well understood. It has been postulated that the meristems are free from virus infection or that the viruses are inactivated by certain components in the culture media.

The success of meristem/shoot tip and bud cultures may depend on the season in which the explants were obtained. Mellor and Stace-Smith (1969) reported that potato buds taken in spring and early summer gave a larger proportion of plantlets than those taken in the other seasons of the year. In a study of micrografting of apple meristem, Huang and Milliakan (1980) reported that, when meristems collected from field-grown plants during November to March were used as scions for grafting, only 10% of the grafted meristems developed into plantlets. On the other hand, when the meristems were taken in May, the rate of success increased to 70%, and then decreased about 10% per month from June to October. These differences in the success of establishing cultures are apparently due to dormancy of the meristems, as evidenced by the work of Altman and Goren (1974) and Tabachnik and Kester (1977). Altman and Goren (1974) demonstrated that the dormancy and the sprouting periods of citrus buds in *in vitro* culture correspond to the natural periods under field conditions. Tabachnik and Kester (1977) found that shoot tips of almond excised from buds in later October, at resting stage, were not able to develop into plantlets. However, if they were stored at 3°C for one to two months, shoot tips were able to grow to plantlets. Buds collected from the fields in late December and January could grow readily, and the shoot development rate was greater in materials collected in January. The cold treatment for materials collected in October thus had similar effects to those of natural vernalization in winter.

Also, it was observed that buds of yams obtained during the active-growth stage can readily develop into plantlets (Ng, unpublished). However, if they were obtained when the plants were beginning to senesce, the regeneration rate of these buds into plantlets was either very low or zero.

Based on the above information, it is clear that actively growing shoot tips/node cuttings should be used for culturing. They can be obtained by growing stem cuttings, bulbs, corms and tubers in sterile soil in the greenhouse or laboratory (Kartha *et al.*, 1974; Wang, 1977; Frison and Ng, 1981) or from actively growing plants in the field. Temperate plants, such as apple, which need cold temperatures for vernalization, should be collected after winter or the buds should be subjected to cold treatment before use. For those plants that have a definite dormant period, the bud should be used only at the end of the dormancy period.

The location from which the bud was collected from the plant may also influence the success of the cultures.

Working on carnation nodal cultures, Roest and Bokelmann (1981) observed that the percentages of shoot development of explants taken from the top and those taken from the base of the same shoot were 88.6% and 69.8%, respectively. Hollings and Stone (1968) reported that meristems excised from the terminal bud of *Chrysanthemum* had a higher success rate than those obtained from lateral buds. Work on cassava has also shown that the apical buds have higher potential to develop into plantlets than the lateral buds (Ng and Hahn, 1985). Of meristems obtained from apical buds, 86.7% developed into plantlets, whereas the highest proportion of lateral buds that developed into plantlets was only 40%. The lateral buds of cassava are usually dormant. They not only have a lower success rate in developing into plantlets, but are also difficult to dissect. Generally, it appears that apical buds of most plant species are more desirable for use in *in vitro* culture. However, because the number of apical buds of a plant is very limited, many workers also use lateral buds (Ancora *et al.*, 1981; Frison and Ng, 1981; Gupta *et al.*, 1981; McComb and Newton, 1981).

Jullien (1983) reported that cells isolated from different tissues of *Asparagus* at different stages of growth responded differently to reduced-temperature storage. Jullien (1983) observed that cells obtained from young cladophylls can be stored better and that the viability of *Asparagus* callus cultures initiated from hypocotyls, roots and stems at two months after storage at 4°C were one-half, three-quarters and one-quarter that of the control (fresh culture), respectively.

2.3.2 Surface disinfection

Surfaces of plant parts carry a wide range of microbial contaminants. To avoid this source of infection, the plant materials must be disinfected before culturing.

Successful removal of surface contaminants from the explant materials is an important factor that determines the establishment of the cultures. Shoot tips and nodal cuttings are usually cut into sizes that are slightly larger than final explants. After surface disinfection with appropriate disinfectants, they are rinsed three times with sterile distilled water to remove any trace amount of disinfectant and they are then cut or dissected under aseptic conditions to the final size and transferred to culture media.

The most commonly used disinfectant is sodium hypochlorite, NaOCl, which is often used as a 5–10% commercial bleach such as Clorox. Other disinfectants, such as hydrogen peroxide, bromine water and mercuric chloride, are also used. It is important to realize that a surface disinfectant is also toxic to the plant tissue. Therefore, the concentration of the disinfecting agents and the duration of treatment should be chosen to minimize tissue death. The concentration of sodium hypochlorite used and the period of disinfection depend on the degree of surface contamination on the explant. Normally, the concentration of sodium hypochlorite can be anywhere between 0.2% and 10%, and the time exposure to the disinfectants between 5 and 20 minutes. For example, sodium hypochlorite at 5% concentration was used to disinfect the shoot tip of rose (Shirvin and Chu, 1979), and 10% was used to disinfect axillary buds of Chinese cabbage (Kuo and Tsay, 1977). However, there are exceptions. For instance, Kunisaki (1980) reported that immersion of *Anthurium* buds in 0.26% sodium hypochlorite for as long as 45 minutes was optimum.

To improve the wettability, a small amount (about 0.1%) of surfactant such as Tween 20, Tween 80 or Teepol was added to the disinfectant. Agitation of the explant and disinfectant mixtures could also increase the efficiency of disinfection.

Treatment with a quick wash of explant materials in 95% ethanol (Vuylsteke and De Langhe, 1985) or 70% ethanol for few minutes prior to immersion in sodium hypochlorite solution has also been used (McComb and Newton, 1981; Frison and Ng, 1981). This treatment not only acts as an effective surfactant, but also kills some of the micro-organisms. In some cases, plant materials were rinsed with running tap water overnight before excision of the buds for disinfection (Kuo and Tsay, 1977).

A two-step disinfection method was found to be effective in removing the contaminants in caladium, taro and cocoyam (Hartmans, 1974). In this case, shoots of about 1.5 cm^3 each with a bud were rinsed in running water, soaked in 0.52% sodium hypochlorite for ten minutes, trimmed further, soaked for another five minutes in 0.26% sodium hypochlorite, and rinsed briefly with sterile distilled water. Each shoot was then dissected to the desired size and transferred to culture medium. The same operation was also successfully used in taro (Jackson *et al.*, 1977) and in *Anthurium* (Kunisaki, 1980).

To minimize contamination, the mother (donor) plants can be kept in pots on a moist-sand bench instead of using the overhead watering (Walkey and Cooper, 1976) or they can be planted in sterile soil in a growth cabinet or greenhouse (Frison and Ng, 1981). Collecting plant materials when the donor plant is at the active-growth stage can also minimize surface contamination. Litz and Conover (1978) observed that there was a slight seasonal fluctuation in contamination rate in papaya apical tissues.

The rate of contamination was highest during the dry winter months, when the growth of plants in the field was very slow.

Despite all of these disinfection efforts, contamination may still persist in some plant materials, particularly when plant materials are obtained from the field. A number of pretreatments may be used to enhance the elimination of contamination. These normally involve watering the donor plant with antibiotic and fungicide mixtures several weeks before shoot tips/buds are excised, or soaking plant materials in a mixture of antibiotics and fungicides for several hours prior to the normal disinfection process. The latter process was found to be very effective in removing bacterial and fungal contamination from nodal cuttings of several *Manihot* spp. (Ng, unpublished). One may even consider incorporating antibiotics in the culture media. Phillips *et al.* (1981) succeeded in controlling contamination, especially bacterial contamination, by incorporating 50 µg/ml of rifampicin in culture media of *Helianthus tubersosus* culture. They reported that the presence of rifampicin had no adverse effect on the rates of cell division, tracheary-element differentiation and DNA synthesis in the culture explant. Nevertheless, one must bear in mind that some antibiotics might cause phytotoxicity at certain concentrations.

2.3.3. Culture medium

Under normal conditions, a culture medium would be designed for sustaining optimal growth of the cultures. However, when reductions in the growth rate are desired, the approach is completely different.

The medium used depends on the size of the explant. A smaller explant will most likely require a more complicated medium than will a bigger explant. A commonly used formulation is Murashige and Skoog's (MS) medium (Murashige and Skoog, 1962) or a modification of this medium. When a small explant is used, i.e. meristems and shoot tips, the initiation medium is designed to support optimal growth. The medium usually contains a basal salt mixture, sucrose, vitamins, inositol and growth hormones (auxin, cytokinin and gibberellic acid). The subsequent subculture can then be cultured on either normal or reduced-growth media, depending on the requirements. Larger initial explants (node cuttings) can be cultured on a normal culture medium or directly on reduced-growth culture medium.

A reduced-growth medium is formulated to allow the explant to develop and grow at a slow rate. This is usually accomplished by using a medium with lower salt concentrations, such as White's medium (White, 1954) or Heller's medium (Heller, 1953), by reducing the strength of normal MS basal medium, by increasing or decreasing the sucrose concentration or by the addition of osmotica (such as mannitol and sorbitol) and growth retardants in the culture medium. The plantlet storage duration can also be prolonged by increasing the volume of the culture medium and storing the plantlet in a bigger culture vessel. The sealing of the culture vessels with parafilm strips or tape can effectively minimize evaporation and avoid culture-medium desiccation (Mullin and Schlegel, 1976; Frison and Ng, 1981).

When one-half strength MS medium without sucrose but with 1 μM indole butyric acid was used, plantlets of coffee were only 3–4 cm in height after two years of storage (Kartha *et al.*, 1981).

White's medium (a low-salt medium) was able to maintain *Xanthosoma sagittifolium* plantlets for up to 80 weeks (Acheampong and Henshaw, 1984) and it was also successfully used to maintain carrot callus cultures for more than 116 weeks without loss of regeneration capability (Reinert and Backs, 1968).

The effects of sucrose concentration and mineral salt supply on the growth of tomato and carnation plantlets were studied by Schnapp and Preece (1986). The medium that gave maximum growth reduction contained 5% sucrose and three-quarters strength MS mineral salts for tomato plantlets, and 5% sucrose and one-quarter to one-half strength MS mineral salts for carnation.

Ng (unpublished) carried out studies on the effects of sucrose and mannitol on plant growth and storage of yam (*Dioscorea alata*). Three

Figure 2.1 Node cuttings of yam ready for inoculation.

different criteria – plantlet establishment percentage, plant height and recovery rate – were used to assess the observations. The explant materials used in this study were node cuttings (Figure 2.1) from previous cultures. The media used were MS medium with 3% and 5% sucrose alone or in combination with 3% mannitol as well as kinetin (0.5 mg/l) or BAP (1 mg/l). The cultures were incubated at 26–20°C with a 16 h photoperiod. Plantlet establishment percentage was recorded two months after culturing, and plant heights were measured seven months after storage (Figure 2.2). At the end of storage (13 months), the nodes from each treatment were transferred to fresh culture medium, which consisted of MS medium with 3% sucrose and 0.5 mg/l kinetin, and incubated under the same culture conditions to assess the recovery rate. The recovery rate is calcu-

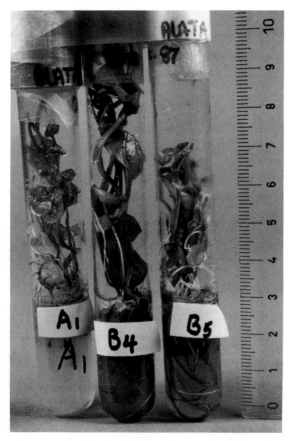

Figure 2.2 Height of yam plantlets grown in different culture media (from left 3% sucrose + kinetin; 5% sucrose + kinetin; 5% sucrose + 3% mannitol + kinetin).

lated according to the number of nodes cultured and the numbers that
showed growth one month after having been retrieved from storage.

The results of this experiment are shown in Table 2.2. Among the six
treatments, the highest plantlet establishment rate was with media III, IV
and V. The supplements for these media were: medium III, 3% sucrose +
3% mannitol + 0.5 mg/l kinetin; medium IV, 5% sucrose + 3% mannitol
+ 0.5 mg/l kinetin; medium V, 3% sucrose + 1 mg/l BAP. The plant growth
was least with medium IV. The recovery rate at the end of the experiment
(13 months) was highest with medium III (100%), followed by medium IV
(90%). The other four treatments had recovery rates of 60% or below.
Based on these criteria, media III and IV (both contain 3% mannitol) were
selected as the favourable media for yam germplasm storage. In medium
IV, though, the plant height at seven months was lower than with medium
III. Nevertheless, the recovery rate with medium IV was lower than that
with medium III. Mannitol has also been used in the conservation of some
other crop species.

2.3.4 Incubation conditions

(a) Reduced incubation temperature
The reduction of incubation temperature has been shown to be very
effective in prolonging the subculturing cycle by reducing the growth rate.
The reduction of incubation light intensity or total darkness in conjunction
with low incubation temperature can also effectively slow down the
growth rate. A normal incubation temperature ranges from 20°C–30°C.
However, temperatures below 25°C could already reduce the growth of
certain crop species. Nevertheless, the recommended temperature regimes

Table 2.2 Effects of culture media on yam growth and storage at normal incubation temperature

Media combination (MS medium with)	Plantlet established (%)	Plant height (cm)	Recovery rate (%)
I. 3% sucrose + K*	80	4.5	60
II. 5% sucrose + K	80	7.7	40
III. 3% sucrose + 3% mannitol + K	100	3.9	100
IV. 5% sucrose + 3% mannitol + K	100	3.3	90
V. 3% sucrose + BAP†	100	7.0	60
VI. 5% sucrose + BAP	60	4.4	50

*Kinetin at 0.5 mg/l.
†Benzyl adenine purine (BAP) at 1 mg/l.

differ from crop to crop. Some crops are more cold-tolerant than others, and the cultures can be maintained at very low temperatures.

Cultures of crop plants were successfully stored at temperatures ranging from 1°C–22°C. Attempts to utilize temperatures below freezing were not successful except in the case of *Prunus* spp. and *Chrysanthemum morifolium*, where plantlets and callus cultures were able to survive at −3°C under dark conditions for more than ten months (Marino *et al.*, 1985) and 52 days (Bannier and Steponkus, 1972), respectively.

Galzy (1969) stored grape shoots in solid media at 9°C for ten months. In liquid media with a filter paper bridge under dark conditions at a temperature of 1°C or 4°C, Mullin and Schlegel (1976) succeeded in storing plantlets derived from meristems of strawberry for six years with the addition of one to two drops of culture media when required. However,

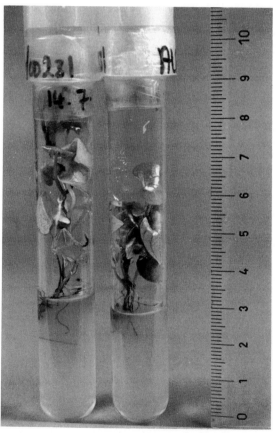

Figure 2.3 Yam plantlet incubated at reduced temperature (16–20°C) for six months (right) and one year (left).

they mentioned that the plantlets were about 0.6 cm in height with roots at least 1.3 cm long before they were transferred to low temperature for storage.

Lundergan and Janick (1979) reported that shoot cultures of apple stored at −17°C did not survive after thawing. Those that were kept at 1°C and 4°C were still alive when stored for over 12 months. Only one cultivar out of 40 was alive by the end of 12 months of storage at 26°C. These high losses were due to a combination of contamination, desiccation and nutrient depletion under normal incubation conditions.

In sugar beet, Miedema (1982) was able to maintain plantlets at 5–10°C with low light intensity for one year. However, he pointed out that the shoot must have had well-developed roots before being transferred to cold storage. He also succeeded in storing flower buds *in vitro* at 5°C and with low light intensity for 13 months. During storage, the flower buds developed into shoots that were yellow in colour. Upon being transferred to 20°C and higher light intensity, the shoots turned green and rooted.

Pregrowth of the cultures under normal incubation conditions to allow the explants to establish before being transferred to lower temperatures for storage was also reported to be important for apple (Lundergan and Janick, 1979), *Prunus* spp. (Marino *et al.*, 1985) and sweet potato, yams and cocoyams (Ng and Hahn, 1985).

Banerjee and De Langhe (1985) were able to store cultivars of banana and plantain at 15°C with 1000 lux light intensity for 12–15 months. Temperatures below 15°C caused rapid deterioration and death of the culture. They observed that the storage period varied, depending on the genomic configuration of the cultivars.

Cheyne and Dale (1980) reported successful storage of temperate forage legume plantlets at 2–6°C with 300 lux light intensity or at 4–6°C but in complete darkness for 15–18 months. The eye grass and orchard grass plantlets were also able to survive at 2–4°C and 300 lux light intensity for 12 months (Dale, 1980).

Zandvoort and Staritsky (1986) reported that they were able to store *Colocasia* at 9°C for three years and *Xanthosoma* spp. at 13°C. They indicated that the addition of growth retardants did not significantly increase the storage duration. Ng and Hahn (1985) demonstrated that yam plantlets can be maintained at a temperature of 16–20°C for over one year (Figure 2.3), and cocoyams can be maintained at the same temperature for one year (Ng, 1986a).

Studying the effects of culture type, proliferating shoot culture and single-shoot rooted culture on the storage of *Vitis* spp. at reduced temperature (9°C), Barlass and Skene (1983) found that different genotypes responded differently. However, the overall results tend to favour single-shoot rooted cultures for the amenable genotypes. The majority of the genotypes stored well for nine to 12 months at 9°C under dark or low-light conditions. Barlass and Skene (1983) confirmed that all plants regenerated after cold storage were phenotypically indistinguishable from the original plants.

Under normal incubation conditions, callus and cell-suspension cultures deteriorate rapidly. The cultures have to be subcultured very often in order to maintain their mitotic and regenerating capability. It has been very difficult to maintain these types of culture.

Using mesophyll cells isolated from *Asparagus officinalis*, Jullien (1983) reported that the suspension culture can be stored at 4°C for one month without altering the growth behaviour of the cultures.

Hiraoka and Kodama (1984) studied the storage of callus cultures of nine plant species (see Table 2.1) at reduced temperature and reported that these species can be stored at 4°C for three to six months. Organogenesis ability (root-forming ability) of the callus cultures stored at 4°C for one year was retained. However, some metabolic products, such as betalain, anthocyanin and shikonin derivatives, were affected. Nevertheless, pre-growth of the cultures at 25°C before subjecting them to cold storage (4°C) was necessary.

A combination of the reduction of incubation temperature and the use of reduced-growth culture media has also given satisfactory results, and indeed was found to be superior in some cases than each of the methods applied alone. One should try to find the best combination for each plant species to be adopted for conservation.

At the IITA we have successfully applied a reduced-growth culture medium in combination with reduced storage temperature (18–20°C) to maintain over 1000 clones of sweet potato germplasm. The culture medium used is the MS salt mixture with vitamins, 3% sucrose, 3% mannitol and 0.6% agar. Nodal cuttings of sweet potatoes were incubated in the medium at normal culturing temperature (25°C) until they developed into plantlets. The established plantlets were then stored at 18–20°C for a period of up to two years (Ng and Hahn, 1985).

It was also reported that potato plantlets could be stored in *in vitro* for two years at 10°C in a low-sucrose culture medium that contained *N*-dimethyl succinamic acid (Mix, 1982).

By using the MS medium at one-quarter strength, Monette (1986) was able to maintain kiwi fruit shoot cultures at 8°C for over one year. He also found that kiwi fruit shoot cultures could survive at 4°C, but the survival rate of the cultures declined substantially after 24 weeks of storage and none of the cultures could survive after 52 weeks of storage. In contrast, 100% of the cultures survived at 8°C after the same period of storage.

A stepwise decrease in incubation temperature of a culture perhaps could enable the culture to be stored at a much lower temperature and could extend the storage duration. This was evident from the work on callus cultures of *Chrysanthemum morifolium* by Bannier and Steponkus (1972). They incubated the callus cultures at 27°C for ten days, and then transferred the cultures to a medium with higher sucrose content and stored them at 4.5°C for two weeks. After this, the cultures were stored at −3.5°C, and they could be stored at this subzero temperature for 52 days. The encouraging results of this study suggested that there may be a good possibility of applying the procedure of the stepwise decrease in incubation temperature to the storage of culture materials at a lower temperature, thus increasing their storage life. Further studies to investigate the possibility of using this technique may be worthwhile.

(b) Modification of gaseous environment
Alteration of the gaseous composition in the atmosphere surrounding the cultures will influence their growth and development. Carbon dioxide and oxygen are the two essential gases for the metabolism and respiration of plants. Alteration of the concentration of either gas should have some effect on the growth of the cultures. This could be applied to reduced-growth storage for germplasm conservation. However, not many studies in this area have been reported. The few that have been reported in the literature are reviewed here.

Bridgen and Staby (1981,1983) applied a low-pressure system and a low-oxygen system to study the effect of these two systems on the growth of shoot and callus cultures of *Chrysanthemum* and tobacco. In their experiments, they found that the growth rates of shoot and callus cultures of both *Chrysanthemum* and tobacco were significantly slower when partial oxygen pressure was below 50 mmHg; this was accomplished either by low atmospheric pressure or by low oxygen pressure, as compared to controls. They concluded that partial oxygen pressure plays an important role in reducing the growth of the cultures. The study on carrot by Kessell and Carr (1972) also showed that the concentration of oxygen was very critical to the growth and differentiation of the tissues of this crop.

In a study on carrot callus cultures, Caplin (1959) cast a mineral oil overlay of up to 6.2 mm thickness over the cultures. He found that the growth of the cultures with mineral oil overlay was slower than that of the control. It is apparent that the mineral oil overlay reduced the oxygen supply to the culture. In grape cultures, Caplin showed that the cultures with the mineral oil overlay kept under low light conditions for 120 days still remained small, green and healthy-looking. Those cultures without a mineral oil overlay turned brown (they had apparently died). An additional advantage of using a mineral oil overlay is that it can minimize water loss from the agar medium.

Using this method, Caplin also reported that more than 30 different strains of plant tissue cultures have been maintained by subculturing at intervals of 3–5 months, over a period of more than three years. There were no apparent changes in growth characteristics after transferred to normal growth conditions.

Modification of the gaseous environment has not been fully exploited in reduced-growth storage of germplasm. Further studies to investigate its potential for reduced-growth storage of plant tissue cultures may be useful.

2.4 PROCEDURE FOR REDUCED-GROWTH STORAGE

The procedure for using meristems and node cuttings as explant materials for the reduced-growth storage of plant germplasm is diagrammatically shown in Figure 2.4. The sources of germplasm described here are either from the field or from *in vitro* cultures.

2.4.1 Preparation of plant materials

Apical buds and node cuttings containing buds are obtained from very actively growing plants from either the field, the greenhouse or the growth chamber. Those that are obtained from plants growing in the greenhouse or growth chamber should have less microbial contamination than those from the field. In both cases, plant materials must be disinfected before they can be used for culturing; *in vitro* cultures can be used directly for subculturing without disinfection.

Surface disinfection is accomplished by soaking plant materials in 70% or 95% ethanol for few minutes or few seconds respectively, followed by soaking in sodium hypochlorite solution (ranging from 0.26% to 10%) with a few drops of Tween 20 for 15–20 minutes. After the treatment with sodium hypochlorite the plant materials are rinsed three times with sterile distilled water in a laminar-flow transfer cabinet. The plant materials are ready for dissection or culturing.

2.4.2 Culture medium

Depending on the types of explant material and on the storage method used, the medium or media used will be different. There are two types of culture medium with regard to storage method: normal-growth culture medium and reduced-growth culture medium. A normal-growth culture medium would be a medium that is designed for supporting optimal growth of the explant materials. One should consult the literature to choose the most suitable medium. Very often a slight modification of the medium is required. The medium used for meristem cultures will be

31

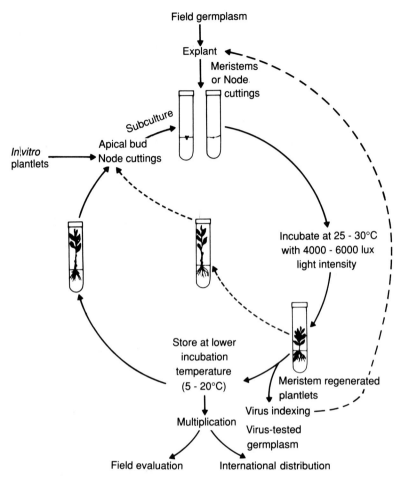

Figure 2.4 Procedure for *in vitro* reduced-growth storage of germplasm.

different from the one used for node cuttings; the medium requirements for meristem cultures are often more complicated than those for node cuttings.

A normal culture medium would consist of a mineral salt mixture, such as that in MS salt mixture, sucrose from 2% to 3%, vitamins and growth hormones (auxin, cytokinin and gibberellic acid).

A reduced-growth medium is aiming at slowing down the growth of the cultures. Therefore, the concentration of mineral salt mixture in the medium can be reduced, sucrose concentration can be increased or decreased, and mannitol or sorbitol (at 3%) and growth retardants (such as abscisic acid and *N*-dimethyl succinamic acid) can be added to the culture medium. The pH of the medium ranges from 5.5 to 6.0. If solid medium is used, agar (0.5–0.7%) is added to the medium and dissolved by heating the medium solution. The medium is then dispensed to test tubes and sterilized in an autoclave at 121°C for 15 minutes.

2.4.3 Dissecting and culturing

The apical buds and lateral buds from the node cuttings are dissected under a stereo microscope with the aid of a sterile scalpel and needle in a laminar-flow transfer cabinet. Meristems with one to two leaf primordia are excised and placed onto a normal growth medium using the tip of the needle or the blade.

Apical buds and node cuttings can also be directly placed on the normal-growth or reduced-growth culture medium after blotting with sterile filter paper to remove the excess water. Usually ten tubes, each with one explant, are prepared for each accession.

The culture tubes are sealed with parafilm strips and labelled. Cultures are incubated at a normal incubation temperature, 25–30°C, with 4000–6000 lux light intensity for culture initiation. After one to two weeks of incubation, cultures are checked visually for possible microbial contamination.

2.4.4 Plant storage

When the regenerated plantlets have shoots of about 1–2 cm in height, they can be kept under normal incubation conditions or transferred to reduced temperature (5–20°C) for storage.

The storage temperature, especially the reduced temperature, varies among crop species and should be studied before being applied on a routine basis. Chilling injuries could affect tropical crops if the temperature was lower than 15°C.

Cultures under storage should be checked periodically to avoid losses due to contamination, media depletion and desiccation. Plantlets that show severe senescence should be subcultured using node cuttings onto fresh culture media, and the storage cycle should be repeated (see Figure 2.4). Theoretically, the storage cycle can be repeated indefinitely, provided that explants retain their regeneration capacity. After subculturing, the old cultures should be kept in the culture room until they are replaced by the newly established cultures.

If all the cultures are contaminated during storage, the plantlets should be either transplanted to soil and brought back into culture when appropriate, or subcultured onto fresh culture medium containing antibiotics to rescue the accession.

Germplasm maintained at normal incubation temperatures or reduced temperatures can be easily retrieved when needed. The node cuttings can be subcultured onto rapid-multiplication media. Plantlets can then be transplated to the field for evaluation. Those that have been virus-tested can be used for international distribution.

2.4.5 Virus indexing

Not all the plants regenerated from meristem cultures are automatically free from virus(es) if the mother plant has been infected. All plants must be thoroughly tested for freedom from virus(es). When the regenerated plantlets attain a height of 3–4 cm with well developed root system, they can be transplanted to sterile soil in an insect-free isolation room for virus indexing. Virus-tested plants can then be put into cultures using node cuttings (see Figure 2.4). The virus-tested plantlets can be used for storage or for international distribution.

Methods used for virus indexing depend on the species concerned and the facilities available. It is advisable to apply as many testing methods as possible to ensure the virus-free status of the regenerated plants. Methods used for indexing include: grafting to test plants, inoculation of leaf sap to test plants, electron microscopic observation of the leaf-sap preparations, serological tests, enzyme-linked immunosorbent assay (Clark and Adams, 1977), serologically specific electron microscopy (Derrick, 1973), electrophoresis and, more recently, recombinant DNA technology.

2.4.6 Monitoring the genetic stability

The genetic stability of plants that are stored *in vitro* is of primary importance. Though it is generally agreed that the meristem is genetically more stable than any other type of explant material, there is evidence that plants regenerated from meristems change their phenotype and these changes are normally chimeras. Efforts should be directed to minimize the genetic instability; in particular, callus formation should be avoided during the culture process. The genetic stability of the stored germplasm should be monitored periodically. The germplasm can be examined phenotypically, cytologically (chromosome number and DNA content) and biochemically (e.g. electrophoresis of isoenzymes and protein analysis) to find whether any genetic changes have occurred.

2.5 CONCLUSIONS

Reduced-growth storage has been used successfully to store cultures (shoot and plantlet) in the short and medium terms. Among the four methods of reduced-growth storage, reduced-incubation-temperature storage has been widely applied to a number of temperate crops as well as tropical crops. The manipulation of culture media has also been used to some extent. Nevertheless, the combination of both has been used to maintain large collections of root and tuber crop germplasm for the past seven to eight years. The method of modification of gaseous environments has been applied to a lesser extent, probably due to its complexity. In the future, microtubers may offer an additional opportunity for storage of germplasm. By manipulating culture media, *in vitro* plantlets could be induced to form microtubers. This has been reported for *Dioscorea alata* (Ammirato, 1984), *D. bulbifera* (Forsyth and Van Staden, 1984), *D. rotundata* (Ng, 1986b), *D. floribunda* (Sengupta *et al.*, 1984) and potato (Estrada *et al.*, 1986).

The genetic instability of the maintained germplasm should not be overlooked. To minimize the genetic changes during storage, it is recommended that meristem/bud culture should be used and that callus formation should be avoided during the regeneration process. The storage cultures should also be periodically monitored for their possible genetic instability.

REFERENCES

Acheampong, E. and Henshaw, G. G. (1984) *In vitro* methods for cocoyams improvement. *Proceedings of the Second Triennial Symposium of the International Society for Tropical Root Crops – Afria Branch*, Doula, Cameroon, pp. 165–168.

Alan, J. J. (1979) Tissue culture storage of sweet potato germplasm. PhD Thesis, University of Birmingham.

Altman, A. and Goren, R. (1974) Growth and dormancy cycles in citrus bud cultures and their hormonal control. *Physiol. Plant.*, **30**, 240–245.

Ammirato, P. V. (1984) in *Handbook of Plant Cell Culture*, Vol. 3, *Crop Species* (ed. P. V. Ammirato, W. R. Sharp and Y. Yamada), Macmillan, New York, pp. 327–354.

Ancora, G., Belli-Donini, M. L. and Cuozzo, L. (1981) Globe artichoke plants obtained from shoot apices through rapid *in vitro* micropropagation. *Sci. Hort.*, **14**, 207–213.

Atanassov, A. I. (1986) in *Biotechnology in Agriculture and Forestry 2. Crops I* (ed. Y. P. S. Bajaj), Springer-Verlag, Heidelberg, Berlin, pp. 462–470.

Banerjee, N. and De Langhe, E. (1985) A tissue culture technique for rapid clonal propagation and storage under minimal growth conditions of *Musa* (Banana and Plantain). *Plant Cell Rep.*, **4**, 351–354.

Bannier, L. J. and Steponkus, P. L. (1972) Freeze preservation of callus cultures of *Chrysanthemum morifolium* Ramat. *HortScience*, **7**, 194.

Barlass, M. and Skene, K. G. M. (1983) Long-term storage of grape *in vitro*. *Plant Genetic Resource Newsletter*, **53**, 19–21.

Bhojwani, S. S. (1981) A tissue culture method for propagation and low temperature storage of *Trifolium repens* genotypes. *Physiol. Plant.*, **52**, 187–190.

Borgman, C. A. and Mudge, K. W. (1986) Factors affecting the establishment and maintenance of 'Titan' red raspberry root organ cultures. *Plant Cell, Tissue and Organ Culture*, **6**(2), 127–137.

Bridgen, M. P. and Staby, G. L. (1981) Low pressure and low oxygen storage of *Nicotiana tabacum* and *Chrysanthemum morifolium* tissue cultures. *Plant Sci. Lett.*, **22**(2), 177–186.

Bridgen, M. P. and Staby, G. L. (1983) in *Handbook of Plant Cell Culture*, Vol. 1, *Technique for Propagation and Breeding* (ed. D. A. Evans, W. R. Sharp, P. V. Ammirato and Y. Yamada), Macmillan, New York, pp. 816–827.

Caplin, S. M. (1959) Mineral oil overlay for conservation of plant tissue cultures. *Am. J. Bot.*, **46**(5), 324–329.

Centro Internacional de Agricultura Tropical (1981) *Annual Report 1980. Tissue Culture, Cassava Program*, CIAT, Cali, Colombia, pp. 79–85.

Cheyne, V. A. and Dale, P. J. (1980) Shoot tip culture in forage legumes. *Plant Sci. Lett.*, **19**, 303–309.

Chun, Y. W. and Hall, R. B. (1986) Low temperature storage of *in vitro* cultured hybrid *Poplar, Populus alba* × *P. grandidentata* plantlets. *VI International Congress of Plant Tissue and Cell Culture*, University of Minnesota, USA, p. 13.

Clark, M. F. and Adams, A. N. (1977) Characteristics of the microplate method of enzyme-linked immunosorbent assay for the detection of plant viruses. *J. Gen. Virol.*, **34**, 475–483.

Dale, P. J. (1980) A method for *in vitro* storage of *in vitro* apple shoots. *HortScience*, **14**, 514.

Derrick, K. S. (1973) Quantitative assay for plant viruses using serologically specific electron microscopy. *Virology*, **56**, 652–653.

Espinoza, N., Estrada, R., Tovar, P., Bryan, J. and Dodds, J. H. (1984) *Tissue culture micropropagation, conservation and export of potato germplasm*, Specialized Technology Document 1, International Potato Center, Lima, Peru, p. 19.

Estrada, R., Tovar, P. and Dodds, J. H. (1986) Induction of *in vitro* tubers in a broad range of potato genotypes. *Plant Cell, Tissue and Organ Culture*, **7**(1), 3–10.

References

Forsyth, C. and Van Staden, J. (1984) Tuberization of *Dioscorea bulbifera* stem nodes in culture. *J. Plant Physiol.*, **115**, 79–83.

Frison, E. A. and Ng, S. Y. (1981) Elimination of sweet potato virus disease agents by meristem tip culture. *Tropical Pest Management*, **27**(4), 452–454.

Galzy, R. (1969) Recherches sur la croissional de *Vitis rupestris* Scheele sain et court noue cultive *in vitro* a differentes temperatures. *Ann. Phytopathol.*, **1**, 149–166.

Gray, D. J. and Mortensen, J. A. (1987) Initiation and maintenance of long term somatic embryogenesis from anthers and ovaries of *Vitis longii* 'Microsperma'. *Plant Cell, Tissue and Organ Culture*, **9**(1), 73–80.

Gupta, P. K., Mascarenhas, A. F. and Jagannathan, V. (1981) Tissue culture of forest trees – clonal propagation of mature trees of *Eucalyptus citriodora* Hook. by tissue culture. *Plant Sci. Lett.*, **20**, 195–201.

Hartmans, R. D. (1974) Dasheen mosaic virus and other phytopathogens eliminated from caladium, taro and cocoyam by culture of shoot tips. *Phytopathology*, **64**, 237–240.

Heller, R. (1953) Research in the mineral nutrition of vegetable tissue culture *in vitro*. *Annales des Sciences Naturelles Botaniques et Biologie Vegetale*, **14**, 1–223.

Hiraoka, N. and Kodama, T. (1984) Effect of non-frozen cold storage on the growth, organogenesis and secondary metabolism of callus cultures. *Plant Cell, Tissue and Organ Culture*, **3**(4), 349–357.

Hollings, M. and Stone, O. M. (1968) Techniques and problems in the production of virus-tested plant material. *Sci. Hort.*, **20**, 57–72.

Huang, S. C. and Milliakan, D. F. (1980) *In vitro* micrografting of apple shoot tips. *HortScience*, **15**(6), 741–743.

Hussey, G. and Hepher, A. (1978) Clonal propagation of sugar beet plant and the formation of polyploids by tissue culture. *Ann. Bot.*, **42**, 477–479.

Jackson, G. V. H., Ball, E. A. and Arditti, J. (1977) Tissue culture of taro, *Colocasia esculenta* (L.) Schott. *J. Hort. Sci.*, **52**, 373–382.

Jarret, R. L. and Fernandez, R. Z. (1984) Shoot-tip vanilla culture for storage and exchange. *Plant Genetic Resources Newsletter*, **57**, 25–27.

Jullien, M. (1983) Medium-term preservation of mesophyll cells isolated from *Asparagus officinalis* L. Development of a simple method by storage at reduced temperature. *Plant Cell, Tissue and Organ Culture*, **2**(4), 305–316.

Kartha, K. K., Constabel, F. and Shyluk, J. P. (1974) Regeneration of cassava plants from apical meristems. *Plant Sci. Lett.*, **2**, 107–113.

Kartha, K. K., Mroginski, L. A., Pahl, K. and Leung, N. L. (1981) Germplasm preservation of coffee (*Coffea arabica* L.) by *in vitro* culture of shoot apical meristem. *Plant Sci. Lett.*, **22**(4), 301–307.

Kessell, R. H. J. and Carr, A. H. (1972) The effect of dissolved oxygen concentration on growth and differentiation of carrot (*Daucus carota*) tissue. *J. Exp. Bot.*, **23**(77), 996–1007.

Kunisaki, J. T. (1980) *In vitro* propagation of *Anthurium andreanum* Lind. *HortScience*, **15**(4), 508–509.

Kuo, C. G. and Tsay, J. S. (1977) Propagation of Chinese cabbage by axillary bud culture. *HortScience*, **12**(5), 456–457.

Litz, R. E. and Conover, R. A. (1978) *In vitro* propagation of papaya. *HortScience*, **13**(3), 241–242.

Lundergan, C. and Janick, J. (1979) Low temperature storage of *in vitro* apple shoots. *HortScience*, **14**(4), 514.

Marino, G., Rosati, P. and Sagrati, F. (1985) Storage of *in vitro* cultures of *Prunus* rootstocks. *Plant Cell, Tissue and Organ Culture*, **5**(1), 73–78.

McComb, J. A. and Newton, S. (1981) Propagation of Kangaroo paws using tissue culture. *J. Hort. Sci.*, **56**(2), 181–183.

Mellor, F. and Stace-Smith, R. (1969) Development of excised potato buds in nutrient culture. *Can. J. Bot.*, **47**, 1617–1621.

Miedema, P. (1982) A tissue culture technique for vegetative propagation and low temperature preservation of *Beta vulgaris*. *Euphytica*, **31**, 635–643.

Mix, G. (1982) *In vitro* preservation of potato materials. *Plant Genetic Resources Newsletter*, **51**, 6–8.

Monette, P. L. (1986) Cold storage of kiwi fruit shoot tips *in vitro*. *Hort-Science*, **21**(5), 1203–1205.

Mullin, R. H. and Schlegel, D. E. (1976) Cold storage maintenance of strawberry meristem plantlets. *HortScience*, **11**(2), 100–101.

Murashige, T. and Skoog, F. (1962) Revised medium for rapid growth and bio-assays with tobacco tissue cultures. *Physiol. Plant.*, **15**, 473–497.

Ng, S. Y. C. (1986a) Application of tissue culture in root crops conservation. *13th Annual Conference of the Genetic Society of Nigeria*, Forestry Research Institute of Nigeria, Ibadan, Nigeria.

Ng, S. Y. C. (1986b) *In vitro* tuberization in *Dioscorea rotundata* Poir (white yam) – A means for germplasm exchange and propagation. *VI International Congress of Plant Tissue and Cell Culture*, University of Minnesota, Minn., USA, p. 255.

Ng, S. Y. C. and Hahn, S. K. (1985) Application of tissue culture to tuber crops at IITA. *Proceedings of the Inter-Centre Seminar on International Agricultural Research Centers and Biotechnology*, The International Rice Research Institute, Los Banos, Philippines, pp. 27–40.

Phillips, R., Arnott, S. M. and Kaplan, S. E. (1981) Antibiotics in plant tissue culture: Rifampicin effectively controls bacterial contaminations without affecting the growth of short-term explant cultures of *Helianthus tuberosus*. *Plant Sci. Lett.*, **21**(3), 235–240.

References

Reinert, J. and Backs, D. (1968) Control of totipotency in plant cells growing *in vitro*. *Nature*, **220**, 1340–1341.

Roberts, E. H. and Street, H. E. (1955) The continuous culture of excised rye roots. *Physiol. Plant.*, **8**, 238–262.

Roest, S. and Bokelmann, G. S. (1981) Vegetative propagation of carnation *in vitro* through multiple shoot development. *Sci. Lett.*, **14**, 357–366.

Schnapp, R. S. and Preece, J. E. (1986) *In vitro* growth reduction of tomato and carnation microplants. *Plant Cell, Tissue and Organ Culture*, **6**(1), 3–8.

Sengupta, J., Mitra, G. C. and Sharma, A. K. (1984) Organogenesis and tuberization in cultures of *Dioscorea floribunda*. *Plant Cell, Tissue and Organ Culture*, **3**(4), 325–331.

Shirvin, R. M. and Chu, M. C. (1979) *In vitro* propagation of 'Forever Yours' rose. *HortScience*, **14**(5), 608–610.

Staritsky, G. (1980) *In vitro* storage of aroid germplasm. *Plant Genetic Resources Newsletter*, **42**, 25–27.

Tabachnik, L. and Kester, D. E. (1977) Shoot culture for Almond and Almond–peach hybrid clones *in vitro*. *HortScience*, **12**(6), 545–547.

Tomes, D. T. (1979) A tissue culture procedure for propagation and maintenance of *Lotus corniculatas* genotypes. *Can. J. Bot.*, **57**, 137–140.

Vuylsteke, D. and De Langhe, E. (1985) Feasibility of *in vitro* propagation of bananas and plantains. *Trop. Agric. (Trinidad)*, **62**(4), 323–328.

Walkey, D. G. A. and Cooper, J. (1976) Growth of *Stellaria media*, *Capsella bursa-pasteris* and *Senecio vulgaris* plantlets from cultured meristem-tips. *Plant Sci. Lett.*, **7**, 179–186.

Wang, P. J. (1977) in *Plant Tissue Culture and its Bio-technological Application* (ed. W. Barz, E. Reinhard and M. H. Zenke), Springer-Verlag, Berlin and Heidelberg, 386–391.

Westcott, R. J., Henshaw, G. G. and Roca, W. M. (1977) Tissue culture storage of potato germplasm: culture initiation and plant regeneration. *Plant Sci. Lett.*, **9**, 309–315.

White, P. R. (1954) *Cultivation of Animal and Plant Cells*, Ronald Press Co., New York, USA.

Withers, L. A. (1982) *Institutes working on tissue culture for genetic conservation. A revised list*, IBPGR Secretariat, Rome, p. 104.

Zandvoort, E. A. and Staritsky, G. (1986) *In vitro* genebanks of tropical aroids – Research of storage conditions. *VI International Congress of Plant Tissue and Cell Culture*, University of Minnesota, Minn., USA, p. 426.

3

Cryopreservation

L. E. TOWILL

3.1 INTRODUCTION

Cryopreservation is a term of recent derivation and refers to the placing and holding of biological materials at low temperatures in a manner such that viability is retained after thawing. The preservation of viability is crucial for plant germplasm conservation because these materials will be used for future plant improvement. Inherent in cryopreservation is the ability to store cells for long periods of time without change or further loss of viability.

Most biological materials that tolerate extensive desiccation, such as some seeds and pollens, are not injured when dried and then put directly into low temperatures, e.g. within liquid nitrogen (LN) (Styles *et al.*, 1982; Stanwood, 1985; Towill, 1985; Akihama and Omura, 1986). However, many propagules do not tolerate extensive drying. Thus, the main problem associated with the low-temperature exposure of these materials is ice formation and the conditions created during this process. The injuring event and how it can be ameliorated through sample pretreatment, application of selected compounds (cryoprotectants), controlled cooling and warming, and post-thaw treatments, is poorly understood. Our knowledge of the cryopreservation of plant protoplasts, cells, tissues and organs is largely empirical and has been derived from observing growth or other measures of viability after application of a particular protocol (Steponkus, 1985). Nevertheless, survival after exposure to liquid nitrogen has been observed in many microbial, animal and plant systems (Kartha, 1985a; Calcott, 1986; Doyle *et al.*, 1988).

This chapter summarizes some observations on plant cryobiology, particularly for isolated shoot tips and buds, and describes some of the studies needed to implement practical cryopreservation. Other related reviews are

In Vitro *Methods for Conservation of Plant Genetic Resources. Edited by John H. Dodds. Published in 1991 by Chapman and Hall, London. ISBN 0 412 33870 X*

by Finkle and Ulrich (1983), Kartha (1985b), Withers (1985a,1987), Sakai (1984,1986) and Chen and Kartha (1987).

3.2 GERMPLASM CONSERVATION

3.2.1 Seed and clonal maintenance

Ample literature is available on the need to conserve plant germplasm for future breeding purposes (Holden and Williams, 1984; Clark and Roath, 1989). The extent of resources allocated to collection, evaluation, characterization and maintenance differs among the diverse crops being conserved. Whatever the extent of the germplasm within gene banks, effective maintenance is required to assure its safety.

Germplasm of many crops is stored in the form of seed. Seeds are grouped according to their desiccation response. Orthodox (desiccation-resistant) seeds retain viability after drying to low moisture contents (~4–7% water on a fresh weight basis), whereas recalcitrant (desiccation-sensitive) seeds lose viability after drying below a relatively high water content (~20–30% water). Orthodox seeds in the dried state are easily preserved at low temperatures because they lack freezable water. Fortunately, many important crop plants have orthodox seeds. The cryopreservation of orthodox seeds is beyond the scope of this chapter. Information can be found in Styles *et al.* (1982) and Stanwood (1985).

Recalcitrant seeds are produced by many tropical and subtropical plants, some aquatic plants and temperate-zone trees with large seeds (Chin and Roberts, 1980). These seeds are difficult to store because of their sensitivity to drying (and thus to freezing, since desiccation occurs during the freezing process) and their general short-lived nature. Combinations of tissue-culture and cryopreservation techniques may enhance germplasm conservation for these species once suitable methodologies are established. Currently, species with recalcitrant seeds are maintained as clones.

Many crop species are maintained clonally, usually for reasons related to inability to form seeds, recalcitrant seeds, large seed size, long juvenile periods (before flowering occurs), and extreme heterozygosity, where continued crossing might disrupt useful gene combinations (de Langhe, 1984; Towill, 1988a). It should be emphasized that germplasm is used primarily in breeding, and a major concern is to ensure that a propagule is capable of flowering.

3.2.2 Active and base collections

Availability and safety in germplasm maintenance are accomplished through the establishment of two types of gene banks (collections). The active gene bank contains materials for ready distribution and evaluation. The base collection is a duplicate set of materials maintained in case loss occurs from the active gene bank. Materials are not extensively distributed from base gene banks and it is desirable that storage be at a different location from that of the active one. This gene bank concept is applicable to both seed and clonal germplasm.

The value of cryopreservation lies in the ability to hold materials for extended periods of time without genetic change, such that periodic regrowth is not needed. This reduces labour costs and avoids potential loss that might occur with continual growth or seed grow-outs. In view of our current knowledge, cryopreservation techniques are more applicable to base-collection maintenance. Although considerable discussion has taken place on the potential application of cryopreservation for germplasm conservation, no clonal gene bank routinely maintains materials in this fashion.

3.2.3 Diversity

Many crops are maintained clonally. The major species held at the US Fruit and Nut Clonal Repositories are given in Table 3.1. These clonal crops encompass tropical, subtropical and temperate species and require vastly different climates for growth. Worldwide, many other important species are held clonally. A few examples include foods (arrowroot, cassava, cocoyam, yam), fibre species (hemp, henequeen, sisal), and spices (cardamom, garlic, ginger, pepper, turmeric). Species may be herbaceous or woody, be chill-sensitive or chill-resistant, and have vastly different tolerances to desiccation and freezing. This diversity in species and in their

Table 3.1 Major crops maintained clonally at different repositories within the United States National Plant Germplasm System

Location	Species maintained clonally
Corvallis, Oregon	Pear, filbert, mint, strawberry, raspberry, blackberry, currants, hops
Davis, California	*Prunus* spp., *Vitis vinifera* and related species
Geneva, New York	Apple, *Vitis* spp.
Brownwood, Texas	Pecans, walnuts
Hilo, Hawaii	Guava, passion fruit, papaya, pineapple, macadamia, lychee, carambola
Griffin, Georgia	Sweet potato
Miami, Florida/Mayaguez, Puerto Rico	Avocado, banana, plantain, coffee, cocao, mango, sugar cane
Riverside, California	*Citrus* spp. and related species, date palm
Orlando, Florida	*Citrus* spp. and related species

attributes must be addressed when devising suitable cryopreservation protocols.

3.3 CRYOPRESERVATION

3.3.1 Systems

Vegetative meristem tips, shoot tips and buds faithfully regenerate the clone upon culture. Procedures for meristem and shoot-tip culture have been described for many herbaceous and some woody species (Styer and Chin, 1983; Hu and Wang, 1983; Kartha, 1985b; Bonga, 1987; Hutchinson and Zimmerman, 1987). Thus, it is logical that the cryopreservation of shoot tips and buds would be useful for germplasm conservation of many clonal species. Whole twigs can be used for cold-hardy species (Sakai and Nishiyama, 1978).

The preservation of species producing recalcitrant seeds could be accomplished through shoot-tip cryopreservation or by direct cryopreservation of the seed or of the excised embryo. To date, cryopreservation of whole seeds by application of cryoprotectants and controlled cooling and warming as described below has had limited success.

The cryobiology of suspension-cultured cells, callus and protoplasts has been described for several species (reviewed by Morris (1980), Finkle and Ulrich (1983), Withers (1985a,b,1987) and Seitz (1987)). These systems are undesirable for germplasm preservation, since a meristem must be regenerated *de novo*. Adventitious shoot formation may increase the likelihood for production of somaclonal variants. Preservation must maintain each clone as true to type as possible and not needlessly introduce variation. Studies with cells and protoplasts have furnished information which may be applied to the cryopreservation of more complex systems, such as shoot tips and buds.

Cryopreservation of suspension cultures and callus which form somatic embryos or secondary products has been reported and has utilized methodology similar to that described below. Recovered lines often retain the original attributes of the parental stock, although only a portion of cells survives freezing (Dougall and Whitten, 1980; Volkova *et al.*, 1987; Seitz and Reinhard, 1987). Embryogenic cell lines have retained the ability to generate somatic embryos after storage in LN (Diettrich *et al.*, 1985; Gupta *et al.*, 1987; Engelmann and Dereuddre, 1988a; Kartha *et al.*, 1988).

3.3.2 Viability assays

Viability tests are needed to assess the effectiveness of the cryogenic protocol. Growth, membrane permeability and metabolic functions are

Table 3.2 Viability assays for plant materials

Function	Viability test
Growth	Resume growth
	Resume normal development
	Cell suspension and callus growth
Membrane integrity	Plasmolysis, deplasmolysis
	Dye exclusion (Evans blue)
	Formazan production and concentration (tetrazolium salt reduction)
	Fluorescein production and concentration from fluorescein diacetate
	Leachate analysis (conductivity proteins/amino acids)
	Multiple freezing points/differential thermal analysis
Currents	Electrical impedance of tissue
Cytoplasmic function	Respiration
	Fixation of labelled amino acids into protein
	Macromolecular syntheses
	Electron transport chain activity
Visual estimates	Water infiltration
	Turgor

usually examined (Table 3.2). Some assays are more useful for whole plants, and woody or leaf tissues, and others for protoplasts and cells. Many researchers have realized 'there is no single method . . . which can be used as an unequivocal criterion of viability' (Palta *et al.*, 1978). As a consequence, two or more assays should be used in confirming comparisons among treatments.

Growth analysis is a common method used to assess viability after treatment. Cells, tissues or shoot tips are placed on an appropriate medium, and growth is measured over a period of time. Comparison of growth among treatments suggests which treatment allows better survival, but quantitative estimates of cell viability are difficult. Growth is the only viability assay that is now readily applied to individual small shoot tips, for instance those derived from *in vitro* plants. It is obvious that adequate growth media must be developed for the species or line of concern.

Various dye assays, such as 2,3,5-triphenyl tetrazolium chloride (Steponkus and Lanphear, 1967; Towill and Mazur, 1975) and fluorescein diacetate (Widholm, 1972), and mixtures, such as Hoechst 33258 and acridine orange (Singh and Stephens, 1986) or fluorescein diacetate and propidium iodide (Huang *et al.*, 1986), are useful for estimating viability in cell and callus cultures. Technically, these are difficult to apply to the evaluation of multicellular units. Conductivity tests of leachates from low-temperature-treated samples are generally not useful for small shoot tips, but are commonly used in cold-acclimation studies for leaf sections or woody tissues and may be modified for cell cultures (Zhang and Willison, 1987). Increased chemiluminescence of LN-treated cells and shoot tips has been correlated with injury (Benson and Noronha-Dutra, 1988) and may serve as a useful viability assay in addition to defining the role of singlet oxygen in the injury process.

Ethylene, ethane and other volatiles are produced during stress and cell death, and could be measured to assess viability and stability during culture (Harber and Fuchigami, 1986; Benson and Withers, 1988). Since these estimates are non-destructive, repeated measurements of the tissue can be made. Nuclear magnetic resonance and infrared spectroscopy may also be used for non-destructive measurements on biological materials, but these require expensive instrumentation. It is uncertain which metabolites and parameters should be analysed to assess viability. All these assays have potential for application to shoot tips, but as yet little information exists to determine usefulness. Different assays often give somewhat different results, and interpretation is critical. All assays ultimately need to be compared with growth assays.

3.3.3 Cryogenic protocols

Cryogenic protocols may be divided into two main groups. The first involves cryopreservation of species that cold-acclimate substantially (approximately −25°C or less). The second is useful for marginally hardy or non-hardy species and utilizes the application of cryoprotectants to excised shoot or meristem tips, followed by controlled cooling and warming.

(a) Cold-hardy species

Materials exhibiting extensive cold-hardiness are best cryopreserved by a method using the natural ability of the plant to tolerate freezing stress. Twigs are preserved and it is essential that they be gathered in a cold-acclimated state. Hardiness is often naturally induced, but some increase in acclimation may be artificially induced with low temperatures (Sakai and Nishiyama, 1978). The extent of acclimation and moisture content must be determined. Desiccation of the twig may be required for less hardy lines.

Sakai and associates demonstrated that if tissues were appropriately treated at subzero temperatures ('prefreezing') for a period of time, they subsequently survived cooling in LN (Sakai, 1986). For example, twigs are cooled slowly (~2°C/h) to temperatures between −30°C and −40°C, held there for a period of time, usually hours, and then immersed directly in LN. Materials in a very cold-hardy state may require cooling to only −10°C to

−20°C prior to immersion (Yakuwa and Oka, 1988). Slow warming of twigs is accomplished by placing them within a refrigerator. Buds are grafted to an appropriate rootstock for regeneration. When this protocol is followed, there appears to be complete survival of the growing tip such that shoot development occurs directly from the grafted bud. This strategy is discussed by Sakai and Nishiyama (1978), Katano et al. (1983), Moriguchi et al. (1985), Sakai (1986), Tyler and Stushnoff (1988a,b) and Tyler et al. (1988).

Small shoot tips can be dissected from LN-treated buds and placed into in vitro culture (Moriguchi et al., 1985; Yakuwa and Oka, 1988). Less hardy genotypes, or materials insufficiently acclimated, may be handled in this manner. Appropriate media are needed to assure plantlet development and root initiation. Information about acclimation is needed to assure a high survival frequency of the transplants.

(b) Minimally cold-hardy or non-cold-hardy species

Many species are not cold-hardy or do not acclimate sufficiently to be treated as described above. These materials required application of suitable compounds (cryoprotectants) to artificially enhance tolerance to low temperatures. Methods use either relatively slow cooling rates (such as 0.25–1°C/min) or rapid cooling rates obtained by immersion directly in LN or other cold solutions. The protocol below pertains to both types of cooling strategies. All permutations of the basic protocol cannot be reviewed, but some modifications are included for discussion.

(i) Pretreatment of plant

Few studies have defined stock plant treatments that enhance explant survival after LN exposure. Species with some ability to cold-acclimate show better survival if the plant or plantlet is given a cold treatment in vivo (carnation (Seibert and Wetherbee, 1977), Digitalis lanata (Diettrich et al., 1987)) or in vitro (apple (Kuo and Lineberger, 1985), Rubus spp. (Reed, 1988)). Survival of cell suspensions or callus is improved if growth prior to cooling is under slight osmotic stress (~4–6% mannitol or sorbitol) or more extreme osmotic stress (~0.5–1 M sorbitol) (Chen et al., 1984a,b). Pretreatment does not increase survival of cells for all species. Four-day exposure to 6% mannitol or sorbitol did not enhance LN survival of soya bean cells (Pritchard et al., 1986a). Pretreatment of cell cultures is addressed in Withers (1985a,b) and Pritchard et al. (1986a,b,c).

(ii) Shoot-tip isolation

Shoot-tip isolation from in vivo or in vitro plants is performed according to standard techniques. The samples used in most studies are about 0.25–1 mm in length and contain a few leaf primordia. Shoot tips derived from in vivo plants are usually somewhat larger than those from in vitro plants. Manipulation of the micropropagation medium or incubation environment may be necessary to have a suitable *in vitro* plant for shoot-tip isolation. Apical shoot tips and axillary shoot tips from the entire length of the *in vitro* plant from potato and mint responded equally well after LN exposure (Towill, unpublished). However, terminal or axillary shoot tips from the top portion of carnations survived in greater percentages than those from lower portions of the plant (Dereuddre *et al.*, 1988). The biology of the species must be considered when selecting the buds from appropriate portions of the *in vivo* or *in vitro* plant.

(iii) Pretreatment of isolated shoot tips

Pretreatment of the isolated shoot tip may be necessary prior to freezing in order to obtain high levels of viability. Grout and Henshaw (1978) cultured shoot tips of *Solanum goniocalyx* for three days prior to freezing. Towill subsequently confirmed the effectiveness of this technique for several potato (*Solanum tuberosum* gp. *tuberosum*) clones and related species (Towill, 1981a,b,1984). The reason for increased survival after preculture without either osmotic stress or cryoprotectant application is not apparent, and the small size of such shoot tips precludes any extensive biochemical study. Another pretreatment includes addition of low levels of cryo-protectants, usually 4–5% (w/v) dimethylsulphoxide (DMSO), within the solid or liquid preculture medium (Haskins and Kartha, 1980). At these concentrations DMSO is apparently not toxic to the shoot tips, and prolonged exposure may ensure its permeation throughout the shoot tip. Preculture of carnation shoot tips with 0.25–0.75 M sucrose and 5% DMSO enhanced their survival after LN treatment (Dereuddre *et al.*, 1988). Mint shoot tips behave similarly after preculture with either sucrose or mannitol (Towill, unpublished).

(iv) Application of cryoprotectant

Cryoprotectants enhance survival of hydrated tissues after exposure to low temperatures (Finkle *et al.*, 1985). DMSO, ethylene glycol, or propylene glycol are commonly used. The most effective cryoprotectants must permeate the cell, and the above three do so fairly rapidly in many plant cells. Some low molecular weight compounds, such as sucrose, glucose, glycerol, proline, and mannitol, do not permeate or do so very slowly, but are effective if added with a permeating cryoprotectant. High molecular weight compounds, such as polyvinylpyrrolidone (PVP), hydroxyethyl starch (HES) and polyethylene glycol (PEG), have been useful in some animal systems but information on their use for plants is limited. They, too, are more effective if combined with permeating cryoprotectants (Towill, unpublished).

Low molecular weight and high molecular weight cryoprotectants are used at concentrations of 1–2 M and 10% (w/v), respectively. Addition in a stepwise fashion minimizes plasmolysis. The cryoprotectant may be added at lower temperatures if toxicity is a problem. Single-step additions are often effective for shoot tips, and this can be explained by the gradual permeation throughout the shoot tip. Incubation is for 1–2 h before initiating the freezing process. Tests which vary the time of exposure to different concentrations are necessary to define toxicity.

(v) Cooling

The sample is initially held for a short time at a temperature just below the freezing point of the cryoprotectant solution, and the solution is nucleated with a small ice crystal or by briefly touching the outside of the tube with LN-cooled forceps. Induced nucleation is necessary in order to avoid spurious effects due to spontaneous nucleation during cooling with subsequent rapid temperature elevation and drop to bath temperatures (Keefe and Henshaw, 1984; Diller, 1985).

The sample is then cooled. Injury over a range of cooling rates has been attributed to two factors (Mazur, 1984). Slowly cooled cells maintain approximate water potential equilibrium by shrinking in response to the decreasing water-vapour pressure. If cooling is too slow, viability declines due to events ascribed to 'solution effects', such as possible pH changes, macromolecular interactions due to cell shrinkage, and increased electrolyte concentrations. Thus, injury is related to dehydration and the mechanical effects of shrinkage. There may be both a time and temperature dependence for some components of injury during cooling exposure.

If cooling is too rapid, vapour-pressure equilibrium between the cell and the extracellular environment is not maintained and intracellular ice forms. Steponkus *et al.* (1985) suggested that rapid ice propagation within the extracellular solution leads to localized charge separations. This destabilizes the plasma membrane and allows penetration of ice into the cell. Intracellular ice is often lethal, although the reasons for lethality are not known.

Cooling rates that give optimal survival for plant cells, protoplasts and small tissue pieces such as shoot tips, are usually about 0.25–1°C/min. Survival decreases rapidly with increased cooling rates such that little survival occurs at rates greater than 4°C/min. Survival at cooling rates less than 0.25°C/min is not well characterized in plants. Cooling rates of between 0.02°C/min and 0.1°C/min have not appreciably diminished mint shoot-tip viability (Towill, unpublished). These slower rates, while useful for describing the events of injury, are not practical for cryopreservation.

For many plant cells, continued slow cooling below −35°C to −40°C leads to considerable loss of viability. In practice, two-step cooling methods are used for preservation. Samples are slowly cooled to between about −30°C and −40°C and then are immersed in LN. After rapid warming, survival approximates that observed for samples warmed directly from the above temperature.

Viability can also be preserved by rapid cooling rates, for instance by immersion of room-temperature or 0°C samples directly into LN. Maximum survival of carnation shoot tips occurred with cooling rates greater than 50°C/min (Seibert and Wetherbee, 1977). Plunging shoot tips directly into LN (rates ~1000–2000°C/min) gave about 20%, 65% and 85% survival with *S. goniocalyx* (Grout and Henshaw, 1978), *Asparagus officinalis* (Kumu *et al.*, 1983) and *Digitalis lanata* (Diettrich *et al.*, 1987), respectively. Retention of viability using either two-step or rapid cooling is probably due to either vitrification of the cell contents or the production of very small ice crystals within the cell such that lethal injury does not occur. It is difficult to generalize on what rapid rates are required to obtain this state in diverse systems.

Other modifications of freezing protocols have been published. Droplet freezing on aluminium foil with slow cooling (Kartha *et al.*, 1982), and 'dry' freezing with rapid cooling (Withers, 1979) have been successful for some materials that otherwise have proven difficult.

(vi) Storage temperatures

Cryopreservation requires storage below about −130°C, where lack of liquid water, extremely slow diffusion, and low molecular energies make the occurrence of chemical reactions, which could lead to deterioration during storage, very infrequent (Mazur, 1984). Liquid nitrogen is a common cryogen because it is inexpensive and relatively non-reactive. Storage temperatures are either −196°C, the temperature of LN, or about −160°C to −180°C, the temperature of the vapour phase above LN. Unfortunately, there are no data on the extended storage of plant materials under these conditions. Potato and cassava shoot tips have been stored for four years (Bajaj, 1985) and *Digitalis lanata* embryogenic lines for three years (Diettrich *et al.*, 1985).

Adequate safeguards must be incorporated into any storage system to prevent the inadventent loss of LN such that warming of the sample might occur. Warming from about −196°C to about −60°C can result in a loss of viability for materials preserved by either the two-step or rapid-cooling process. This is probably due to either devitrification events or the restructuring of smaller, non-disruptive ice crystals into larger crystals. In either case intracellular ice formation is then lethal.

Storage at warmer temperatures, such as between −20°C and −70°C, is not desirable because of shorter lifetimes. This is easily understood for less hardy materials trated with cryoprotectants, since chemical reactions can still occur at measurable rates. Storage of hardy materials at a low moisture content may be more practical at somewhat warmer temperatures than that of LN, but, again, few data exist to give useful information on possible longevities.

There is little in the literature to suggest whether temperatures lower than −196°C are advantageous for preservation. It is interesting to note that partially dehardened cortical cells of *Cornus sericea* lost viability during cooling from −196°C (LN) to −269°C (liquid helium) (Guy *et al.*, 1986).

(vii) Warming

The effects of warming (and subsequent thawing) have not been studied as extensively as those of cooling. Rapid warming is required to retain viability in plant cells and shoot tips with two-step and rapid-cooling procedures. This avoids ice recrystallization or devitrification within the cells (Macfarlane, 1986), but direct evidence in plant systems is sparse. Warming is accomplished by immersing vials containing the cells or tips into a water bath (~35–40°C) for a few seconds (warming rates about 200–500°C/min depending up on the volume of liquid in the tube). Optimum conditions are not defined because of the technical difficulties in varying warming rates over rather narrow ranges.

Cold-hardy twigs cooled slowly and transferred to LN have higher viabilities if warmed slowly. But warming is often performed by merely placing materials within a cold room, and actual warming rates for the tissues are not reported.

(viii) Post-thawing treatment

Post-thaw treatment of samples has also received little attention. Samples are diluted directly with liquid growth medium to remove the cryo-protectant(s). This somewhat rapid dilution may be detrimental to some cells (Withers, 1979). Slower diffusion of the cryoprotectant from the sample is achieved by placing the sample on agar and reculturing to fresh medium a short time later. However, rate of dilution of LN-treated shoot tips from mint and potato did not affect viability (Towill, unpublished).

Sugar cane and rice callus showed greater growth if dilution was performed at 22°C compared to 0°C (Finkle and Ulrich, 1982). LN-treated shoot tips of mint gave a similar response (Towill, unpublished). It is not known why higher temperatures and slower dilution rates are less injuri-ous, but the interaction with membrane phase transitions occurring at the lower temperatures may be important. Rates of cryoprotectant efflux from cells or tissues have not been determined under the stated conditions.

Selection of a suitable growth medium and culture environment for treated shoot tips is important. Growth regulator concentrations may require adjustment in order to enhance survival and growth of the shoot tip (Grout *et al.*, 1978; Towill, 1981b; Withers *et al.*, 1988). However, callus production must be avoided to reduce the frequency of somaclonal variants after shoot initiation (Henshaw *et al.*, 1985a; Withers *et al.*, 1988).

(ix) Other materials

Methods for treatment of hydrated whole seeds, dissected portions of seeds and somatic embryos are similar to those described for shoot tips.

Hydrated seeds have been proposed as models for recalcitrant seeds. Rapid cooling rates can allow seeds, for example lettuce at 19% moisture, to survive LN exposure (Roos and Stanwood, 1981). At greater moisture levels cryoprotectants are required. Hydrated tomato seeds (49% moisture) survived cooling at 12°C/min or 850°C/min if exposed to 35% DMSO (Grout and Crisp, 1985). Maize seeds, hydrated to about 24% moisture, were viable after rapid freezing when treated with 5 M 1,2-propanediol (de Boucaud and Cambecedes, 1988). Inclusion of sucrose or sucrose and DMSO enhanced survival. Recalcitrant seeds from *Anthurium* survived LN exposure if treated with high concentrations of some cryoprotectants, but results were not consistent (P. Stanwood, personal communication). Since the plasma membrane is a major site of freezing injury (Steponkus, 1984), it is uncertain if hydrated, but otherwise orthodox, seeds are really analogous to recalcitrant seeds. The sensitivity to desiccation suggests that recalcitrant seeds have a different membrane composition, so mere hydration may not provide an analogous system to study interactions.

Some seeds which are difficult to preserve or recalcitrant may withstand low temperatures without cryoprotectants if embryonic axes are dissected out and carefully lowered in moisture content before LN treatment. This was demonstrated with difficult-to-preserve seeds of *Elaesis guinensis* L. (Grout *et al.*, 1983) and with true recalcitrant seeds of *Araucaria hunsteinii* (Pritchard and Prendergast, 1986) and *Hevea brasiliensis* (Normah *et al.*, 1986). In the latter case, embryonic axes desiccated and containing 14–20% moisture survived immersion in LN and subsequently developed normally *in vitro*. Partial desiccation may also explain the observations with *Anthurium* (P. Stanwood, personal communication). Coconut embryos treated with cryoprotectants have also been reported to survive LN treatment (Bajaj, 1984).

Somatic embryos and synthetic seeds, if orthodox, could be used for germplasm preservation of clonal lines because manipulation, storage and regeneration may be simpler than the tedious excision, pretreatment and cryopreservation of shoot tips or buds. Somatic embryogenesis must be induced in all genotypes, and in mature plants for woody species. The frequency of somaclonal variants produced from somatic embryos should be low to avoid possible selection of off-types during subsequent propagation. Plants derived from somatic embryos may show a lower frequency of variation than those produced by organogenesis from other tissue-culture systems (Vasil, 1986; Maheswaran and Williams, 1987), but conclusive data for diverse species are lacking. To be suitable for storage, somatic embryos should exhibit desiccation tolerance and show longevities similar to those of sexually derived seeds. Viability after drying is often low but this may be due to the occurrence in the population of somatic embryos in different developmental stages, insufficient drying rates, or lack of optimization in any of the several steps leading to a mature somatic embryo (Gray, 1987; Gray *et al.*, 1987).

Survival of hydrated somatic embryos after LN exposure has been reported (Bajaj, 1978; Engelmann *et al.*, 1985; Bertrand-Desbrunais *et al.*, 1988; Engelmann and Dereuddre, 1988a,b; Marin and Duran-Vila, 1988). Samples were not dried and cryoprotectants were necessary to obtain survival; however, survival was often low. Little information exists on the long-term, low-temperature storability of somatic embryos or synthetic seeds.

3.3.4 Freezing injury and theories of cryoprotection

A thorough discussion of freezing injury and of possible mechanisms by which cryoprotectants function is beyond the scope of this chapter. Several recent reviews examine current thoughts on these subjects (Mazur, 1984; Steponkus, 1984,1985; Morris and Clarke, 1987; Singh and Laroche, 1988).

Various lines of evidence implicate membrane damage, particularly to the plasma membrane, as a primary component of freezing injury (Steponkus, 1984). Some lipid components convert from a liquid-crystalline to a gel phase during cooling and may not reconstitute during warming (Quinn, 1985). Cryoprotectants apparently protect membranes from damage during cooling and aid in protecting proteins and nucleic acids from inactiviation. Studies with anhydrobiotic organisms and propagules have helped to elucidate mechanisms of protein and membrane protection during severe desiccation (Caffrey *et al.*, 1988; Carpenter and Crowe, 1988; Crowe *et al.*, 1988). Compounds conferring protection are diverse and include amino acids, polyols, sugars and lyotropic salts.

Freezing injury is apparently not just a single event (Mazur, 1984; Steponkus, 1984). Treatments protecting against conditions that may injure at higher temperatures may not afford protection against conditions that may occur at lower temperatures. Thus, some cryoprotectants may enhance survival at somewhat lower temperatures but not at −196°C.

3.4 RESEARCH AND DEVELOPMENT NEEDS

To date, evidence suggests that cryopreservation of shoot tips or buds could be useful for germplasm conservation of clonal species. The implementation of cryopreservation technology requires preliminary studies to define conditions for maximum survival, to evaluate the genetic integrity of retrieved samples, and to test applicability to a range of diversity. Practical aspects also require attention.

Table 3.3 Examples of freeze preservation of meristem or shoot tips using cryo-protectants

Species	Cooling method	Reference
Arachis hypogaea	Rapid, slow	Bajaj (1979)
Asparagus officinalis	Rapid, two-step	Kumu *et al.* (1983)
Beta vulgaris	Two-step	Braun (1988)
Brassica oleracea	Two-step	Harada *et al.* (1985)
Brassica napus	Rapid, two-step	Withers *et al.* (1988)
	Rapid	Benson and Noronha–Dutra (1988)
Cicer arietinum	Rapid, slow	Bajaj (1979)
Dianthus caryophyllus	Two-step	Dereuddre *et al.* (1987)
	Rapid	Seibert and Wetherbee (1977)
Digitalis lanata	Rapid, two-step	Diettrich *et al.* (1987)
Fragaria × *ananassa*	Two-step	Sakai *et al.* (1978)
Fragaria × *ananassa*	Two-step	Kartha *et al.* (1980)
Haplopappus gracilis	Two-step	Taniguchi *et al.* (1988)
Lycopersicon esculentum	Rapid	Grout *et al.* (1978)
Malus domestica	Two-step	Kuo and Lineberger (1985)
Manihot esculenta	Two-step	Kartha *et al.* (1982)
Manihot esculenta	Rapid, two-step	Bajaj (1983,1985)
Mentha spp.	Two-step	Towill (1988b)
Morus bombycis	Two-step	Yakuwa and Oka (1988)
Pisum sativum	Two-step	Kartha *et al.* (1979)
		Haskins and Kartha (1980)
Pyrus serotina	Two-step	Moriguchi *et al.* (1985)
Rubus spp.	Two-step	Reed and Lagerstedt (1987)
		Reed (1988)
Solanum tuberosum	Rapid	Bajaj (1985)
		Henshaw *et al.* (1985a,b)
	Two-step	Bajaj (1981,1983,1985)
		Manzhulin *et al.* (1984)
		Manzhulin (1985)
		Towill (1981a,1983, 1984)
Solanum spp.	Rapid	Grout and Henshaw (1978, 1980)
		Henshaw *et al.* (1985a,b)
	Two-step	Towil (1981a,1984)

3.4.1 Maximizing survival

Shoot tips from various species have survived LN exposure (Table 3.3). Several species have shown survival using a single method (Towill, unpublished) (Table 3.4). Two observations are important. A variable percentage of shoot tips survive cryopreservation, and growth of LN-treated shoot tips is often different from that of non-frozen but cryoprotectant-treated controls.

(a) Growth of treated shoot tips

Cell proliferation in LN-treated pea shoot tips occurred predominantly

Table 3.4 Percentage of shoot tips surviving LN exposure using two-step freezing*

	Treatment	
Species	Control	LN
Beta vulgaris	100 (5/5)†	18 (10/54)
Chrysanthemum sp.	100 (16/16)	81 (25/31)
Dianthus caryophyllus	87 (13/15)	90 (26/29)
Fragraria × *ananassa*	100 (12/12)	64 (25/36)
Mentha aquatica × *spicata*	100 (6/6)	82 (23/28)
Malus domestica	100 (13/13)	78 (11/14)
Phoenix dactylifera	82 (9/11)	63 (10/16)
Raphanus sp.	100 (9/9)	50 (18/36)
Ribes grossularia	100 (22/22)	67 (14/22)
Rubus laciniatus	100 (12/12)	92 (24/26)
Solanum tuberosum	100 (15/15)	44 (18/41)
Spirea sp.	100 (10/11)	23 (10/44)

*Method consists of preculture for two days in 4% DMSO within growth medium, addition of either 8% DMSO + 8% sucrose or 12% DMSO within growth medium (1 h, 22°C), 0.25°C/min cooling to −35°C, immersion in LN, storage for 1–7 days, thawing in a 40°C water bath, dilution of cryoprotectant from sample, and plating on solid growth medium.
†Percentage of shoot tips treated that showed growth. Numbers in parentheses are number of growing shoot tips over the number treated. Controls were exposed to the cryoprotectants but not frozen.

from superficial cells on primordial leaves (Haskins and Kartha, 1980). Shoots quickly regenerated from these proliferations. Shoot tips from potato species and cultivars produced a small amount of callus prior to shoot formation (Grout and Henshaw, 1978; Towill, 1981a,b,1984). Growth-media modifications allowed direct development of potato shoot tips into a shoot (Henshaw *et al.*, 1985b). Mint species produced a small amount of callus (1–2 mm), usually in the meristem dome region, before shoot initiation (Towill, 1988b). *Digitalis* after LN treatment formed 'green nodular tissue' from which shoots formed (Diettrich *et al.*, 1987). However, most *Rubus* shoot tips directly developed into a shoot (Reed and Lagerstedt, 1987).

Thus, differential cell survival often occurs in LN-treated shoot tips. This raises several questions. Do the shoots that rapidly appear in some systems (e.g. mint) originate from axillary meristems within the shoot tip, or are they adventitious? Why do some shoot tips survive LN treatment while others do not? Is survival related to variation in the number of surviving cells, the efficacy of cryoprotectant permeation within the shoot tips, or inherent variability in the shoot-tip population? Rapid cooling of *Brassica napus* shoot tips resulted in higher percentages of regrowth and of shoot regeneration compared to a two-step cooling protocol. This suggests that more cells survived within the rapidly cooled shoot tips (Withers et al., 1988). Unfortunately, comprehensive microscopy data are limited mainly to a few reports for two-step cooling.

The basis for differential cell survival in shoot tips is not known. Animal cells derived from different tissues have different cooling rate optima depending partially on water-permeability properties and cell size/volume ratio (Mazur, 1984). Some cryoprotectants are more toxic to certain animal cells. For plants, most cells examined have similar cooling-rate optima, but cryoprotectant toxicity is not well characterized.

(b) Vitrification

Cell survival may be enhanced by employing conditions favouring vitrification. Vitrification refers to solidification ('glassification') of the system during cooling without ice formation. The concept was initially discussed in the 1930s, but more recently has been proposed for cryopreservation of larger tissue pieces or organs from animals (Fahy et al., 1984,1987). The physical aspects associated with this phenomenon are known (Vassoille and Perez, 1985; Angell and Choi, 1986; Boutron and Mehl, 1986; MacFarlane, 1986,1987). Large concentrations of cryoprotectants and appropriate (usually rapid) cooling and warming rates are required (Fahy et al., 1984,1987). Addition of the cryoprotectant decreases the cooling rate necessary to obtain vitrification by lowering the homogeneous nucleation temperature and increasing viscosity (MacKenzie, 1977). Potential advantages of vitrification strategies include application to larger pieces of tissue, convenience in cooling, and avoiding or minimizing 'solution effects' injury.

Three issues are of practical importance for plant cell and shoot-tip cryopreservation. The appropriate combination of cryoprotectants which induce vitrification but are not too toxic must be determined (Fahy, 1986). The permeation of these compounds into cells and within tissues must be characterized. The potential effectiveness in enhancing survival must be assessed. Very little is known about these issues for plants.

Mixtures of cryoprotectants are empirically derived from their phase behaviour. Sufficient amounts must permeate the cell such that vitrification is possible at fairly high subzero temperatures. It is uncertain what intracellular concentrations are required and what contribution endogenous components play in the vitrification process. Gradual addition of the cryoprotectant appears necessary in order to avoid plasmolysis. However, a balance must be attained, since the large concentrations required are somewhat toxic.

Survival using vitrification has been reported for animal embryos (Rall, 1987) and some animal tissues, including human monocytes (Takahashi et al., 1986) and human islets of Langerhans (Jutte et al., 1987). Vitrification probably accounts for the survival of LN-treated cold-hardy plants (Hirsh *et al.*, 1985; Hirsh, 1987).

(c) Recoverable injury

Are cells irreversibly damaged after low-temperature exposure? Some microbes exhibit sublethal injury, and the recovered fraction can be increased through appropriate post-thawing culture (Mackey, 1984; Ray, 1984). Sublethal injury also occurs in animal systems (McGann *et al.*, 1975; Law *et al.*, 1979,1980), but it is not clear whether this occurs in plant systems. Enhanced survival after plating on different culture media may merely mean that viable cells given appropriate conditions (such as growth-regulator addenda) can now divide. It does not provide evidence of repair of sublethal injury within cells, i.e. of yielding an increased frequency of viable cells. Observations of potassium and calcium fluxes and of viability after freezing signify that sublethal injury occurs, at least after exposure to certain temperature ranges (Arora and Palta, 1988). The questions remain to what extent repair is possible, what the mechanisms are, and whether repair can be practically manipulated to increase survival frequency.

Although the initial site of injury appears to be the plasma membrane, the progression of events that then occur has not been determined. Post-injury events may lead to death, ostensibly through the irreversible loss of membrane semipermeability characteristics. Free-radical production and lipid peroxidations have been observed after injury. Both could lead to a range of reactions within the cell. A fundamental question remains. Are certain events reversible and, if so, can external intervention ameliorate the extent of injury? Elucidating the temporal sequence of metabolic changes that occur after injury is an initial step in determining if injury can be repaired.

3.4.2 Genetic stability

The genetic integrity of samples must be ensured before cryopreservation is accepted for long-term storage of plant germplasm. The major concern is what potential exists for somaclonal variation in these systems? Other concerns include the stability of materials held for long periods of time in LN, and whether the cryogenic treatment, including application of cryoprotectants and cooling/thawing, is mutagenic.

Plants derived from manipulations *in vitro* are occasionally not true to type (Earle and Demarly, 1982; Semal, 1985). However, plants derived from shoot-tip elongation and axillary-bud proliferation *in vitro* are typically normal. Thus, methods are needed to ensure that most cells within the shoot tip survive the low-temperature treatments and that the recovery medium promotes direct growth of the shoot tip. But, as discussed above, the axillary origin of shoots produced after cryogenic treatment is often difficult to determine. Hence, a population of regenerants should be analysed to determine if somaclonal variation is significant. Although the original genotype is probably preponderant in regenerants, propagation from a gene bank would utilize only one to a few individuals. The question then becomes, what are the consequences of repeated selections over time for a gene bank? Would an altered accession be obtained? Change could be minimized by using several regenerants to represent the next clonal generation. This is obviously easier for *in vitro* maintained materials than for those that have been field maintained. Another question is whether inferences drawn from studies using relatively few genotypes are valid for the diversity of the species.

DNA-repair processes are not functioning at low temperatures, yet cells are exposed to background levels of irradiation. Therefore, what are the limits to long-term storage before background irradiation produces genetic change or lethality (Ashwood-Smith and Grant, 1977; Lyon *et al.*, 1977; Ashwood-Smith and Friedmann, 1979; Glenister *et al.*, 1984)? Experimentally, this is difficult to determine. Extrapolations from studies in which animal cells were exposed to high levels of radiation while cryogenically stored suggest that lethality would occur only after several thousand years of exposure to background levels. Storage at low temperatures is clearly not indefinite, but the data suggest that one or more centuries of storage is realistic.

Are cryogenic protocols mutagenic (Calcott and Gargett, 1981; Ashwood-Smith, 1985)? No clear consensus has emerged from studies using microbes, but data suggest that mutation rates do not increase. A related process, freeze-drying, is mutagenic, but mutagenicity is usually attributed to events during drying and not to cooling and thawing *per se* (Ashwood-Smith and Grant, 1976; Tanaka *et al.*, 1979). Some studies suggest that DNA is damaged during cryogenic exposure (Calcott, 1986), but less DNA damage and increased survival occurred in samples treated with cryoprotectants. Studies with eukaryotes are less common, but indicate that chromosome damage is not evident (Ashwood-Smith and Friedmann, 1979; Ashwood-Smith, 1985). Thus, many researchers feel that 'standard cryopreservation procedures are without genetic hazards' (Ashwood-Smith, 1985).

3.4.3 Application to diverse genotypes

Methods developed for cryopreservation have utilized only a few genotypes. Which portions of the cryogenic protocol can be standardized for all or most genotypes and which will require development for specific genotypes? Data for diverse materials hint that the cryogenic portion can be standardized for shoot tips from most species. Many survive a rather similar two-step cooling protocol. This may also be true for rapid cooling but the data are not as extensive.

More work is needed to define optimum cryoprotectant combinations and to develop and assess recovery and growth media. These, too, may be species- or line-specific. Whether more generalized procedures can be developed awaits suitable investigation. Determining whether genotypes respond similarly or differently requires knowledge of survival-frequency variation occurring among experiments. The literature does not generally contain such data, but this variation is occasionally large (Henshaw *et al.*, 1985b; Towill, unpublished). This has obvious implications for practical application to gene banks.

3.4.4 Practical aspects of cryopreservation

A cryogenic-storage system for shoot tips and twigs requires careful consideration of handling, packaging and storage. Some arbitrary decisions must be made based upon available information. For instance, what levels of viability are sufficient to ensure safe preservation of germplasm, how many replicates should be stored, and how many shoot tips are within a replicate? What periodic testing during storage is required to ensure that loss has not occurred? Should samples be grown out periodically to determine if change has occurred? How should change be detected? These are·similar to questions posed for the storage of seeds.

Equipment needs must be considered. What type of cooling bath will be used? Methanol baths cooled by compressors are useful for research, but do not accommodate large numbers of samples. Leakage of methanol into samples may occur. Liquid nitrogen vapour-pulsed cooling units may be more practical, but there must be certainty of nucleating each sample during the cooling stage. What apparatus is appropriate if rapid cooling rates are required, especially when large numbers of samples are involved? Mechanization is preferred to avoid variation introduced by human manipulations.

Packaging for long-term storage is important. In contrast to orthodox seeds, stored samples of shoot tips and twigs cannot be thawed, repackaged and again cooled. Plastic ampules (1–5 ml volumes) are available from several manufacturers and appear to be adequate for small-volume samples. However, if fast cooling is used, other types of vials may be needed to attain the desired rates.

Liquid nitrogen storage refrigerators come in a range of sizes and appear to be adequate for long-term preservation. Alarm systems are necessary in case levels of liquid nitrogen get too low. Retrieval from the refrigerator should be sufficiently easy such that other samples are not inadvertently warmed.

Although cryopreservation is envisioned for base-collection storage, procedures for shipment of samples remain to be determined. Overnight delivery services and well-insulated containers permit shipment in the frozen state. Personnel at the receiving station must be trained to receive and recover samples. Alternatively, samples can be thawed at the gene bank and distributed as *in vitro* plant cultures. This requires additional effort but may result in enhanced viability.

3.5 SUMMARY

Considerable progress has been made over the past 20 years in defining conditions that allow hydrated plant cells to survive low-temperature exposure. A better understanding of anhydrobiosis and freezing injury has provided insight into what is needed to improve viability. The mechanism of cryoprotection, however, remains elusive. Application of this knowledge to gene banks awaits innovations in dealing with diversity.

The questions posed in this chapter have no simple solution. Some are specific to a species. Others require decisions based upon good judgement. Cryopreservation of cold-hardy twigs may now be feasible for gene banks, and development studies should be done. Cryopreservation of less cold-hardy propagules requires more detailed study before extensive application to gene banks is possible.

REFERENCES

Angell, C. A. and Choi, Y. (1986) Crystallization and vitrification in aqueous systems. *J. Microscopy*, **141**, 251–261.

Akihama, T. and Omura, M. (1986) Preservation of fruit tree pollen, *Biotechnology in agriculture and forestry*, **1**. Trees I. (ed. Y. P. S. Bajaj), Springer-Verlag, Berlin, pp. 101–112.

Arora, R. and Palta, J. P. (1988) *In vivo* perturbation of membrane-associated calcium by freeze–thaw stress in onion bulb cells. Simulation of this perturbation in extracellular KCl and alleviation by calcium. *Plant Physiol.*, **87**, 622–628.

Ashwood-Smith, M. J. (1985) Genetic damage is not produced by normal cryopreservation procedures involving either glycerol or DMSO: a cautionary note however on the possible effects of DMSO. *Cryobiology*, **22**, 427–433.

Ashwood-Smith, M. J. and Friedmann, G. B. (1979) Lethal and chromosomal effects of freezing, thawing, storage time and x-irradiation on mammalian cells preserved at −196°C in dimethylsulfoxide. *Cryobiology*, **16**, 132–140.

Ashwood-Smith, M. J. and Grant, E. (1976) Mutation induction in bacteria by freeze-drying. *Cryobiology*, **13**, 206–213.

Ashwood-Smith, M. J. and Grant, E. (1977) in *The Freezing of Mammalian Embryos* (ed. K. Elliot and J. Whelan), Elsevier, Amsterdam, pp. 251–272.

Bajaj, Y. P. S. (1978) Effects of super-low temperature on excised anthers and pollen-embryos of *Atropa*, *Nicotiana* and *Petunia*. *Phytomorphology*, **28**, 171–176.

Bajaj, Y. P. S. (1979) Freeze preservation of meristems of *Arachis hypogaea* and *Cicer arietinum*. *Ind. J. Exp. Biol.*, **17**, 1405–1407.

Bajaj, Y. P. S. (1981) Regeneration of plants from potato meristems freeze-preserved for 24 months. *Euphytica*, **30**, 141–145.

Bajaj, Y. P. S. (1983) Cassava plants from meristem cultures freeze-preserved for three years. *Field Crops Res.*, **7**, 161–167.

Bajaj, Y. P. S. (1984) Induction of growth in frozen embryos of coconut and ovules of *Citrus*. *Curr. Sci.*, **53**, 1215–1216.

Bajaj, Y. P. S. (1985) Cryopreservation of germplasm of potato (*Solanum tuberosum* L.) and cassava (*Manihot esculenta* Crantz): viability of excised meristems cryopreserved up to 4 years. *Ind. J. Exp. Biol.*, **23**, 285–287.

Benson, E. E. and Noronha-Dutra, A. A. (1988) Chemiluminescence in cryopreserved plant tissue cultures: the possible involvement of singlet oxygen in cryoinjury. *Cryo-Letters*, **9**, 120–131.

Benson, E. E. and Withers, L. A. (1988) Gas chromatographic analysis of volatile production by cryopreserved plant tissue cultures: a non-destructive method for assessing stability. *Cryo-Letters*, **8**, 35–46.

Bertrand-Desbrunais, A., Fabre, J., Engelmann, F., Dereuddre, J. and Charrier, A. (1988) Adventive embryogenesis recovery from coffee (*Coffee arabica* L.) somatic embryos after freezing in liquid nitrogen. *C.R. Acad. Sci.*, **307**, 795–816.

Bonga, J. M. (1987) in *Cell and Tissue Culture in Forestry*, Vol. 1 (ed. J. M. Bonga and D. J. Durzan), Martinus Nijhoff, Dordrecht, pp. 249–271.

Boutron, P. and Mehl, P. (1986) Nouveaux cryoprotecteurs pour la cryopreservation des cellules par vitrification totale. *Bull. Soc. Bot. France, Actual Bot.*, **133**, 27–39.

Caffrey, M., Fonseca, V. and Leopold, A. C. (1988) Lipid–sugar interactions. Relevance to anhydrous biology. *Plant Physiol.*, **86**, 754–758.

Calcott, P. H. (1986) Cryopreservation of microorganisms. *CRC Crit. Rev. Biotech.*, **4**, 279–297.

Calcott, P. H. and Gargett, A. M. (1981) Mutagenicity of freezing and thawing. *FEMS Microbiol. Lett.*, **10**, 151–155.

Carpenter, J. F. and Crowe, J. H. (1988) Modes of stabilization of a protein by organic solutes during desiccation. *Cryobiology*, **25**, 459–470.

Chen, T. H. H. and Kartha, K. K. (1987) in *Cell and Tissue Culture in Forestry*, Vol. 2 (ed. J. M. Bonga and D. J. Durzan), Martinus Nijhoff, Dordrecht, pp. 305–319.

Chen, T. H. H., Kartha, K. K., Constabel, F. and Gusta, L. V. (1984a) Freezing characteristics of cultured *Catharanthus roseus* (L.) G. Don cells treated with dimethylsulfoxide and sorbitol in relation to cryopreservation. *Plant Physiol.*, **75**, 720–725.

Chen, T. H. H., Kartha, K. K., Leung, N. L., Kurz, W. G. W., Chatson, K. B. and Constable, F. (1984b) Cryopreservation of alkaloid-producing cell cultures of periwinkle (*Catharanthus roseus*). *Plant Physiol.*, **75**, 726–731.

Chin, H. F. and Roberts, E. H. (1980) *Recalcitrant Crop Seeds*, Tropical Press, Kuala Lumpur.

Clark, R. L. and Roath, W. W. (eds) (1989) *U.S. Plant Germplasm System – Science and Technology* (*Plant Breeding Reviews*, Vol. 7), Timber Press, Portland, Oregon.

Crowe, J. H., Crowe, L. M., Carpenter, J. F., Rudolph, A. S., Wistrom, L. A., Spargo, B. J. and Anchordoguy, T. J. (1988) Interactions of sugars with membranes. *Biochim. Biophys. Acta*, **947**, 367–384.

de Boucaud, M.-T. and Cambecedes, J. (1988) The use of 1,2-propanediol for cryopreservation of recalcitrant seeds: the model case of *Zea mays* imbibed seeds. *Cryo-Letters*, **9**, 94–101.

de Langhe, E. A. L. (1984) in *Crop Genetic Resources: Conservation and Evaluation* (ed. J. H. W. Holden and J. T. Williams), George Allen and Unwin, London, pp. 131–137.

Dereuddre, J., Fabre, J. and Bassaglia, C. (1987) Resistance to freezing in liquid nitrogen of carnation (*Dianthus caryophyllus* L. var Eolo) apical and axillary shoot tips excised from different aged *in vivo* plants. *Plant Cell Rep.*, **7**, 170–173.

Diettrich, B., Haack, U., Popov, A. S., Butenko, R. G. and Luckner, M. (1985) Long-term storage in liquid nitrogen of an embryogenic cell strain of *Digitalis lanata. Biochem. Physiol. Pflanzen*, **180**, 33–43.

Diettrich, B., Wolf, T., Bormann, A., Popov, A. S., Butenko, R. G. and Luckner, M. (1987) Cryopreservation of *Digitalis lanata* shoot tips. *Plant. Med.*, **53**, 359–363.

Diller, K. R. (1985) The influence of controlled ice nucleation on regulating thermal history during freezing. *Cryobiology*, **22**, 268–281.

Dougall, D. K. and Whitten, G. H. (1980) The ability of wild carrot cell cultures to retain their capacity for anthocyanin synthesis after storage at −140°C. *Plant. Med.*, Supplement, pp. 129–135.

Doyle, A., Morris, C. B. and Armitage, W. J. (1988) in *Upstream Processes:*

Equipment and Techniques (ed. A. Mizrahi), Alan R. Liss Inc., New York, pp. 1–17.

Earle, E. D. and Demarly, Y. (eds) (1982) *Variability in Plants Regenerated from Tissue Cultures*, Praeger, New York.

Engelmann, F. and Dereuddre, J. (1988a) Cryopreservation of oil palm somatic embryo: importance of the freezing process. *Cryo-Letters*, **9**, 220–235.

Engelmann, F. and Dereuddre, J. (1988b) Effets du milieu de culture sur la production d'embryoides desints a la cryoconservation chez le palmier a huile (*Elaeis guineensis* Jacq.). *C. R. Acad. Sci.*, **306**, 515–520.

Engelmann, F., Duval, Y. and Dereuddre, J. (1985) Survival and proliferation of oil palm (*Elaeis guineensis* Jacq.) somatic embryos after freezing to liquid nitrogen. *C. R. Acad Sci.*, **301**, 111–116.

Fahy, G. M. (1986) The relevance of cryoprotectant 'toxicity' to cryobiology. *Cryobiology*, **23**, 1–13.

Fahy, G. M., MacFarlane, D. R., Angell, C. A. and Meryman, H. T. (1984) Vitrification as an approach to cryopreservation. *Cryobiology*, **21**, 407–426.

Fahy, G. M., Levy, D. I. and Ali, S. E. (1987) Some emerging principles underlying the physical properties, biological actions and utility of vitrification solutions. *Cryobiology*, **24**, 196–213.

Finkle, B. J. and Ulrich, J. (1982) Cryoprotectant removal temperature as a factor in the survival of frozen rice and sugarcane cells. *Cryobiology*, **19**, 329–335.

Finkle, B. J. and Ulrich, J. (1983) in *Techniques for Propagation and Breeding*, (*Handbook of Plant Cell Culture*, Vol. 1), (ed. D. A. Evans, W. R. Sharpe, P. V. Ammirato and Y. Yamada), Macmillan. New York, pp. 806–815.

Finkle, B. J., Zavala, M. E. and Ulrich, J. M. (1985). in K. K. Kartha *Cryopreservation of Plant Cells and Organs* (ed. K. K. Kartha), CRC Press, Boca Raton, FL, pp. 75–113.

Glenister, P. H., Whittingham, D. G. and Lyon, M. F. (1984) Further studies on the effect of radiation during the storage of frozen 8-cell mouse embryos at −196°C. *J. Reprod. Fertil.*, **70**, 229–234.

Gray, D. J. (1987) Quiescence in monocotyledonous and dicotyledonous somatic embryos induced by dehydration. *HortScience*, **22**, 810–814.

Gray, D. J., Conger, B. V. and Songstad, D. D. (1987) Desiccated quiescent somatic embryos of orchard grass for use as synthetic seeds. *In vitro*, **23**, 29–33.

Grout, B. W. W. (1987) in *The Effects of Low Temperatures on Biological Systems*, (ed. B. W. W. Grout and G. J. Morris), Arnold Press, London, pp. 293–314.

Grout, B. W. W. and Crisp, P. (1985) Germination as an unreliable indicator of the effectiveness of cryopreservative procedures for imbibed seeds. *Ann. Bot.*, **55**, 289–292.

Grout, B. W. W. and Henshaw, G. G. (1978) Freeze preservation of potato shoot tip cultures. *Ann. Bot.*, **42**, 1227–1229.

Grout, B. W. W. and Henshaw, G. G. (1980) Structural observations on the growth of potato shoot-tip cultures after thawing from liquid nitrogen. *Ann. Bot.*, **46**, 243–248.

Grout, B. W. W., Wescott, R. J. and Henshaw, G. G. (1978) Survival of shoot meristems of tomato seedlings frozen in liquid nitrogen. *Cryobiology*, **15**, 478–483.

Grout, B. W. W., Shelton, K. and Pritchard, H. W. (1983) Orthodox behaviour of oil palm seed and cryopreservation of the excised embryo for genetic conservation. *Ann. Bot.*, **52**, 381–384.

Gupta, P. K., Durzan, D. J. and Finkle, B. J. (1987) Somatic polyembryogenesis in embryogenic cell masses of *Picea abies* (Norway spruce) and *Pinus taeda* (loblolly pine) after thawing from liquid nitrogen. *Can. J. For. Res.*, **17**, 1130–1134.

Guy, C. L., Niemi, K. J., Fennell, A. and Carter, J. V. (1986) Survival of *Cornus sericea* L. stem cortical cells following immersion in liquid helium. *Plant, Cell and Environment*, **9**, 447–450.

Harada, T., Inaba, A., Yakuwa, T. and Tamura, T. (1985) Freeze-preservation of apices isolated from small heads of Brussels sprouts. *HortScience*, **20**, 678–680.

Harber, R. M. and Fuchigami, L. H. (1986) The relationship of ethylene and ethane production to tissue damage in frozen Rhododendron leaf discs. *J. Am. Soc. Hort. Sci.*, **111**, 434–436.

Haskins, R. H. and Kartha, K. K. (1980) Freeze preservation of pea meristems: cell survival. *Can. J. Bot.*, **58**, 833–840.

Henshaw, G. G., Keefe, P. D. and O'Hara, J. F. (1985a) in *In Vitro Techniques: Propagation and Long Term Storage* (ed. A. Schafer-Menuhr), Martinus Nijhoff/Dr W. Junk Publ., Dordrecht, pp. 155–160.

Henshaw, G. G., O'Hara, J. F. and Stamp, J. A. (1985b) *Cryopreservation of Plant Cells and Organs* (ed. K. K. Kartha), CRC Press, Boca Raton, FL, pp. 159–170.

Hirsh, A. G. (1987) Vitrification in plants as a natural form of cryoprotection. *Cryobiology*, **24**, 214–228.

Hirsh, A. G., Williams, R. J. and Meryman, H. T. (1985) A novel method of natural cryoprotection. *Plant Physiol.*, **79**, 41–56.

Holden, J. H. W. and Williams, J. T. (eds) (1984) *Crop Genetic Resources. Conservation and Evaluation*, George Allen and Unwin, London.

Hu, C. Y. and Wang, P. J. (1983) *Techniques for Propagation and Breeding* (*Handbook of Plant Cell Culture*, Vol. 1) (ed. D. A. Evans, W. R. Sharp, P. V. Ammirato and Y. Yamada), Macmillan, NY, pp. 177–227.

References

Huang, C.-N., Cornejo, M. J., Bush, D. S. and Jones, R. L. (1986) Estimating viability of plant protoplasts using double and single staining. *Protoplasma*, **135**, 80–87.

Hutchinson, J. F. and Zimmerman, R. H. (1987) Tissue culture of temperate fruit and nut trees. *Hort. Rev.*, **9**, 273–349.

Jutte, N., Heyse, P., Hansen, H. G., Bruining, G. J. and Zeilmaker, G. H. (1987) Vitrification of human islets of Langerhans. *Cryobiology*, **24**, 403–411.

Kartha, K. K. (ed.) (1985a) *Cryopreservation of Plant Cells and Organs*, CRC Press, Boca Raton, Fl.

Kartha, K. K. (1985b) in *Cryopreservation of Plant Cells and Organs* (ed. K. K. Kartha), CRC Press, Boca Raton, Fl, pp. 115–134.

Kartha, K. K., Leung, N. L. and Gamborg, O. L. (1979) Freeze-preservation of pea meristems in liquid nitrogen and subsequent plant regeneration. *Plant Sci. Lett.*, **15**, 7–15.

Kartha, K. K., Leung, N. L. and Pahl, K. (1980) Cryopreservation of strawberry meristems and mass propagation of plantlets. *J. Am. Soc. Hort. Sci.*, **105**, 481–484.

Kartha, K. K., Leung, N. L. and Mroginski, L. A. (1982) *In vitro* growth responses and plant regeneration from cryopreserved meristems of cassava (*Manihot esculenta* Crantz). *Z. Pflanzenphysiol.*, **107**, 133–140.

Kartha, K. K., Fowke, L. C., Leung, N. L., Caswell, K. L. and Hakman, I. (1988) Induction of somatic embryos and plantlets from cryopreserved cell cultures of white spruce (*Picea glauca*). *J. Plant Physiol.*, **132**, 529–539.

Katano, M., Ishihara, A. and Sakai, A. (1983) Survival of dormant apple shoot tips after immersion in liquid nitrogen. *HortScience*, **18**, 707–708.

Keefe, P. D. and Henshaw, G. G. (1984) A note on the multiple role of artificial nucleation of the suspending medium during two-step cryopreservation procedures. *Cryo-Letters*, **5**, 71–78.

Kumu, Y., Harada, T. and Yakuwa, T. (1983) Development of a whole plant from a shoot tip of *Asparagus officinalis* L. frozen down to −196°C. *J. Fac. Agr. Hokkaido Univ.*, **61**, 285–294.

Kuo, C.-C. and Lineberger, R. D. (1985) Survival of *in vitro* cultured tissue of 'Jonathan' apples exposed to −196°C. *HortScience*, **20**, 764–767.

Law, P., Lepock, J. R. and Kruuv, J. (1979) Effect of protective agents on amount and repair of sublethal freeze–thaw damage in mammalian cells. *Cryobiology*, **16**, 430–435.

Law, P., Lepock, J. R. and Kruuv, J. (1980) Effect of temperature and protein synthesis inhibitors on repair of sublethal freeze–thaw damage in mammalian cells. *Cryo-Letters*, **1**, 324–336.

Lyon, M. F., Whittingham, D. G. and Glenister, P. (1977) in *The Freezing of*

Mammalian Embryos (ed. K. Elliot and J. Whelan), Elsevier, Amsterdam, pp. 273–281.

MacFarlane, D. R. (1986) Devitrification in glass-forming aqueous solutions. *Cryobiology*, **23**, 230–244.

MacFarlane, D. R. (1987) Physical aspects of vitrification in aqueous solution. *Cryobiology*, **24**, 181–195.

MacKenzie, A. P. (1977) Non-equilibrium freezing behaviour of aqueous systems. *Philos. Trans R. Soc. Lond. B*, **278**, 167–189.

Mackey, B. M. (1984) in *The revival of Injured Microbes* (ed. M. E. H. Andrew and A. D. Russell), Academic Press, New York, pp. 45–75.

Maheswaran, G. and Williams, E. G. (1987) Uniformity of plants regenerated by direct somatic embryogenesis from zygotic embryos of *Trifolium repens*. *Ann. Bot.*, **59**, 93–98.

Manzhulin, A. V. (1985) Factors affecting survival of potato stem apices after deep-freezing. *Sov. Plant Physiol.* **31**, 500–504.

Manzhulin, A. V., Butenko, R. G. and Popov, A. S. (1984) Effect of preliminary preparation of potato apices on their viability after deep-freezing. *Sov. Plant Physiol.*, **30**, 910–914.

Marin, M. L. and Duran-Vila, N. (1988) Survival of somatic embryos and recovery of plants of sweet orange (*Citrus sinensis* (L.) Osb.) after immersion in liquid nitrogen. *Plant Cell, Tissue and Organ Culture*, **14**, 51–57.

Mazur, P. (1984) Freezing of living cells: mechanisms and implications. *Am. J. Physiol.*, **247**, c.125–142.

McGann, L. E., Kruuv, J., Frim, J. and Frey, H. E. (1975) Factors affecting the repair of sublethal freeze–thaw damage in mammalian cells. I. Suboptimal temperature and hypoxia. *Cryobiology*, **12**, 530–539.

Moriguchi, T., Akihama, T. and Kozaki, I. (1985) Freeze-preservation of dormant pear shoot apices. *Japan. J. Breed.*, **35**, 196–199.

Morris, G. J. (1980) in *Temperature Preservation in Medicine and Biology* (ed. M. J. Ashwood-Smith and J. Farrant), University Park Press, Baltimore, Maryland, pp. 253–284.

Morris, G. J. and Clarke, A. (1987) in *The Effects of Low Temperatures on Biological Systems* (ed. B. W. W. Grout and G. J. Morris), Arnold Press, Baltimore, Maryland, pp. 72–119.

Normah, M. N., Chin, H. F. and Hor, Y. L. (1986) Desiccation and cryopreservation of embryonic axes of *Hevea brasiliensis* Muell.-Arg. *Pertanika*, **9**, 299–303.

Palta, J. P., Levitt, J. and Standelmann, E. J. (1978) Plant viability assay. *Cryobiology*, **15**, 249–255.

Pritchard, H. W. and Prendergast, F. G. (1986) Effects of desiccation and cryopreservation on the *in vitro* viability of embryos of the recalcitrant seed species *Araucaria hunsteinii* K. Schum. *J. Exp. Bot.*, **37**, 1388–1397.

Pritchard, H. W., Grout, B. W. W. and Short, K. C. (1986a) Osmotic stress as a pregrowth procedure for cryopreservation. 1. Growth and ultrastructure of sycamore and soybean cell suspension. *Ann. Bot.*, **57**, 41–48.

Pritchard, H. W., Grout, B. W. W. and Short, K. C. (1986b) Osmotic stress as a pregrowth procedure for cryopreservation. 2. Water relations and metabolic state of sycamore and soybean cell suspension. *Ann. Bot.*, **57**, 371–378.

Pritchard, H. W., Grout, B. W. W. and Short, K. C. (1986c) Osmotic stress as a pregrowth procedure for cryopreservation. 3. Cryobiology of Sycamore, maple and soybean cell suspensions. *Ann. Bot.*, **57**, 379–387.

Quinn, P. J. (1985) A lipid-phase separation model of low-temperature damage to biological membranes. *Cryobiology*, **22**, 128–146.

Rall, W. F. (1987) Factors affecting the survival of mouse embryos cryopreserved by vitrification. *Cryobiology*, **24**, 387–402.

Ray, B. (1984) in *Repairable Lesions in Microorganisms* (ed. A. Hurst and A. Nasim), Academic Press, New York, pp. 237–271.

Reed, B. M. (1988) Cold acclimation as a method to improve survival of cryopreserved *Rubus* meristems. *Cryo-Letters*, **9**, 166–171.

Reed, B. M. and Lagerstedt, H. B. (1987) Freeze preservation of apical meristems of *Rubus* in liquid nitrogen. *HortScience*, **22**, 302–303.

Roos, E. E. and Stanwood, P. C. (1981) Effects of low temperature, cooling rate, and moisture content on seed germination of lettuce. *J. Am. Soc. Hort. Sci.*, **106**, 30–34.

Sakai, A. (1984) Cryopreservation of apical meristems. *Hort. Rev.*, **6**, 357–372.

Sakai, A. (1986) in *Trees I (Biotechnology in Agriculture and Forestry*, Vol. 1), (ed. Y. P. S. Bajaj) Springer-Verlag, Berlin, pp. 113–129.

Sakai, A. and Nishiyama, Y. (1978) Cryopreservation of winter vegetative buds of hardy fruit trees in liquid nitrogen. *HortScience*, 13, 225–227.

Sakai, A., Yamakawa, M., Sakata, D., Harada, T. and Yakuwa, T. (1978) Development of a whole plant from an excised strawberry runner apex frozen to −196°C. *Low Temp. Sci. Er. B*, **36**, 31–38.

Seibert, M. and Wetherbee, P. J. (1977) Increased survival and differentiation of frozen plant organ cultures through cold treatment. *Plant Physiol.*, **59**, 1043–1046.

Seitz, U. (1987) Cryopreservation of plant cell cultures. *Plant. Med.*, **53**, 311–314.

Seitz, U. and Reinhard, E. (1987) Growth and ginsenoside patterns of cryoprotected Panax ginseng cell cultures. *J. Plant Physiol.*, **131**, 215–223.

Semal, J. (ed.) (1985) *Somaclonal Variations and Crop Improvement*, Martinus Nijhoff, Dordrecht.

Singh, J. and Laroche, A. (1988) Freezing tolerance in plants: a biochemical overview. *Biochem. Cell Biol.*, **66**, 650–657.

Singh, N. P. and Stephens, R. E. (1986) A novel technique for viable cell determinations. *Stain Technol.*, **61**, 315–318.

Stanwood, P. C. (1985) in *Cryopreservation of Plant Cells and Organs* (ed. K. K. Kartha), CRC Press, Boca Raton, Fl, pp. 199–226.

Steponkus, P. L. (1984) Role of the plasma membrane in freezing injury and cold acclimation. *Annu. Rev. Plant Physiol.*, **35**, 543–584.

Steponkus, P. L. (1985) in *Biotechnology in Plant Science: Relevance to Agriculture in the Eighties* (ed. M. Zaitlin, P. Day and A. Hollaender), Academic Press, New York, pp. 145–159.

Steponkus, P. L. and Lanphear, F. O. (1967) Refinement of the triphenyl tetrazolium chloride method of determining cold injury. *Plant Physiol.*, **42**, 1423–1426.

Steponkus, P. L., Stout, D. G., Wolfe, J. and Lovelace, R. V. E. (1985) Possible role of transient electric fields in freezing induced membrane destabilization. *J. Membr. Biol.*, **85**, 191–198.

Styer, D. J. and Chin, C. K. (1983) Meristem and shoot-tip culture for propagation, pathogen elimination and germplasm preservation. *Hort. Rev.*, **5**, 221–227.

Styles, E. D., Burgess, J. M., Mason, C. and Huber, B. M. (1982) Storage of seed in liquid nitrogen. *Cryobiology*, **19**, 195–199.

Takahashi, T., Hirsh, A., Erbe, E. F., Bross, J. B., Steere, R. L. and Williams, R. J. (1986) Vitrification of human monocytes. *Cryobiology*, **23**, 103–115.

Tanaka, Y., Yoh, M., Takeda, Y. and Miwane, T. (1979) Induction of mutation in *Escherichia coli* by freeze drying. *Appl. Environ. Microbiol.*, **37**, 369–372.

Taniguchi, K., Tanaka, R., Ashitani, N. and Miyagawa, H. (1988) Freeze preservation of tissue-cultured shoot primordia of the annual *Haplopappus gracilis* ($2n = 4$). *Jpn. J. Genet.*, **63**, 267–272.

Towill, L. E. (1981a) *Solanum etuberosum*: a model for studying the cryobiology of shoot-tips in the tuber-bearing *Solanum* species. *Plant Sci. Lett.*, **20**, 315–324.

Towill, L. E. (1981b) Survival at low temperatures of shoot-tips from cultivars of *Solanum tuberosum* group tuberosum. *Cryo-Letters*, **2**, 373–382.

Towill, L. E. (1983) Improved survival after cryogenic exposure of shoot tips derived from *in vitro* plantlet cultures of potato. *Cryobiology*, **20**, 567–573.

Towill, L. E. (1984) Survival at ultra-low temperatures of shoot tips from *Solanum tuberosum* groups andigena, phureja, stenotomum, tuberosum and other tuber-bearing *Solanum* species. *Cryo-Letters*, **5**, 319–326.

References

Towill, L. E. (1985) in *Cryopreservation of Plant Cells and Organs*, (ed. K. K. Kartha), CRC Press, Boca Raton, Fl, pp. 171–198.

Towill, L. E. (1988a) Genetic considerations for germplasm preservation of clonal materials. *HortScience*, **23**, 91–97.

Towill, L. E. (1988b) Survival of shoot tips from mint species after short-term exposure to cryogenic conditions. *HortScience*, **23**, 839–841.

Towill, L. E. and Mazur, P. (1975) Studies on the reduction of 2,3,5-triphenyltetrazolium chloride as a viability assay for plant tissue cultures. *Can. J. Bot.*, **53**, 1097–1102.

Tyler, N. and Stushnoff, C. (1988a) The effects of prefreezing and controlled dehydration on cryopreservation of dormant vegetative apple buds. *Can. J. Plant Sci.*, **68**, 1163–1167.

Tyler, N. and Stushnoff, C. (1988b) Dehydration of dormant apple buds at different stages of cold acclimation to induce cryopreservability in different cultivars. *Can. J. Plant Sci.*, **68**, 1168–1176.

Tyler, N., Stushnoff, C. and Gusta, L. V. (1988) Freezing of water in dormant vegetative apple buds in relation to cryopreservation. *Plant Physiol.*, **87**, 201–205.

Vasil, I. K. (1986) in *Somaclonal Variations and Crop Improvement* (ed. J. Semal), Martinus Nijhoff, Dordrecht, pp. 108–116.

Vassoille, R. and Perez, J. (1985) Etudes physiques des melanges eau-cryoprotecteurs. *Ann. Physiol. Fr.*, **10**, 307–367.

Volkova, L. A., Gorskaya, N. V., Popov, A. S., Paukov, V. N. and Urmantseva, V. V. (1987) Preservation of the main characteristics of *Dioscorea* mutant cell strains after storage at extremely low temperatures. *Sov. Plant Physiol.*, **33**, 598–605.

Widholm, J. M. (1972) The use of fluorescein diacetate and phenosafranine for determining viability of cultured plant cells. *Stain Technol.*, **47**, 189–194.

Withers, L. A. (1979) Freeze preservation of somatic embryos and clonal plantlets of carrot (*Daucus carota* L.). *Plant Physiol.*, **63**, 460–467.

Withers, L. A. (1985a) in *Cell Culture and Somatic Cell Genetics of Plants*, Vol. 2 (ed. I. K. Vasil), Academic Press, New York, pp. 253–316.

Withers, L. A. (1985b) in *Cryopreservation of Plant Cells and Organs* (ed. K. K. Kartha), CRS Press, Boca Raton, Fl, pp. 243–268.

Withers, L. A. (1987) in The Effects of Low Temperatures on Biological Systems (ed. B. W. W. Grout and G. J. Morris), Arnold Press, London, pp. 389–409.

Withers, L. A., Benson, E. E. and Martin, M. (1988) Cooling rate/culture medium interactions in the survival and structural stability of cryopreserved shoot-tips of *Brassica napus*. *Cryo-Letters*, **9**, 114–119.

Yakuwa, H. and Oka, S. (1988) Plant regeneration through meristem culture from vegetative buds of mulberry (*Morus bombycis* Koidz.) stored in liquid nitrogen. *Ann. Bot.*, **62**, 79–82.

Zavala, M. E. and Sussex, I. M. (1986) Survival of developing wheat embryos and bean axes following cryoprotection and freezing in liquid nitrogen. *J. Plant Physiol*, **122**, 193–197.

Zhang, M. I. N. and Willison, J. H. M. (1987) An improved conductivity method for the measurement of frost hardiness. *Can. J. Bot.*, **65**, 710–715.

4

Molecular analysis of genetic stability

R. H. POTTER and M. G. K. JONES

4.1 INTRODUCTION

Various techniques of tissue culture can be used for propagation, germplasm storage, germplasm transfer or genetic manipulation of crop plants (Jones and Karp, 1985). The methodology and advantages of tissue culture for micropropagation, germplasm conservation and storage are described in more detail elsewhere in this volume.

It is now well established that the tissue-culture regimes to which plant cells are subjected can have a profound effect on the genetic constitution of plants regenerated from culture. Originally, the regeneration of plants from explants or protoplasts via a callus phase was thought simply to be a method of cloning plants, and it was believed that the resultant plants would be genetically identical to the parental tissues. Careful examination of plants regenerated in such a way has now shown unequivocally that this is usually not the case and that such plants may vary morphologically, cytologically and biochemically from the parental material. This phenomenon has been termed 'somaclonal variation', and has been described extensively (Larkin and Scowcroft, 1981; Karp and Bright, 1985). However, the available evidence suggests that plants regenerated following tissue-culture regimes that maintain meristem organization are genetically uniform, and do not exhibit somaclonal variation (Denton *et al.*, 1977; Wright, 1983). Thus micropropagation has been employed extensively, both in research and for commercial purposes, as a method for multiplying plants rapidly and under controlled conditions. To date the genetic uniformity of plants propagated in this way has been assumed rather than rigorously tested (e.g. Hussey and Stacey, 1981; El-Gizawy and Ford-Lloyd, 1987). The question that has been posed on a number of occasions,

In Vitro *Methods for Conservation of Plant Genetic Resources. Edited by John H. Dodds. Published in 1991 by Chapman and Hall, London. ISBN 0 412 33870 X*

when micropropagated plants have been distributed worldwide, is: 'are these plants identical to the parental material?'. The answer given usually takes the form: 'Yes, well, we hope that they are'.

There is clearly a need to examine this question in more detail, so that an unequivocal answer can be given. At present there is very little information in the literature on stability following micropropagation and storage, and as a result we have set out to develop methods that can answer the question, and to apply these to one specific crop plant, potato. Some reports of variation following micropropagation highlight the need to develop efficient methods of genotype fingerprinting, particularly for stock mother cultures used for bulking up specific genotypes.

In the following discussion, methods that can be applied for examination of germplasm stability following micropropagation and storage are outlined, and the progress to date in their application is described.

4.2 CURRENT METHODS OF ASSESSING GENETIC UNIFORMITY

Current methods employed to assess the stability of plant genotypes are outlined in Table 4.1. Morphological characters have been the main criteria by which plant varieties have been identified. At present, crop plants such as potato are essentially classified into varieties on the basis of characters such as growth habit, leaf morphology, flower morphology, tuber shape and colour, although other characteristics (e.g. disease and pest resistance and quality) are also taken into account (National Institute of Agricultural Botany, 1975). Commercial seed-potato producers must rigorously 'rogue' their fields to remove those plants which do not conform to the required morphological characteristics. Whilst such an approach is the simplest method of monitoring functional purity of stocks, there are many aspects of variation that can occur but not be manifested in obvious morphological changes.

One type of biochemical character that can be examined is protein composition (Denton et al., 1977). Soluble proteins extracted from different genotypes or tissue sources can be compared by techniques based on electrophoresis. Gel electrophoresis can be carried out either under conditions that denature the proteins or conditions that do not. The spectrum of proteins present in an extract can be separated and compared on the basis of size (molecular weight), charge and conformation. A higher resolution of protein patterns can be obtained by two-step ('two-dimensional') electrophoresis, using different buffer systems, or by using isoelectric focusing in the first dimension, followed by gel electrophoresis in the second. The latter approach improves resolution and separation of

72

Table 4.1 Methods of assessing stability

Phenotypic	– Morphology	– quantitative, e.g. height	
		qualitative, e.g. flower colour	
	– Protein electrophoresis	– denaturing	
		non-denaturing	
		isoelectric focusing	
		non-specific stain	
		enzyme activity	
		immunological staining	
	– Secondary products, e.g. alkaloid production		
		gaseous evolution	
Genetic	– Chromosomes	– general staining	– aneuploidy
		giemsa/C-banding	– inversions
			deletions
	– Restriction fragment analysis		
		– alterations in DNA sequence	

similar proteins. The proteins are then visualized by staining. Usually this involves Coomassie blue, or silver stains (for total proteins), and specific proteins (e.g. isoenzymes) can be identified by activity-related stains, or by using specific antibodies.

The approaches outlined above have been used extensively, not only to assess stability and uniformity, but also to look for evidence of somaclonal variation. They have also been used routinely to identify varieties (Stegeman and Schnick, 1985) and as a test for quality of agricultural produce.

Another method of biochemical analysis that has been investigated as a marker for stability is analysis of various classes of secondary metabolites. This approach has been used particularly in connection with cryopreservation (Benson and Withers, 1987), and industrial plant cell culture (Fowler, 1984).

Although such methods have proved to be very useful, they do not provide a complete picture of genetic stability. This is because they are based on phenotypic characters, or the expressed products of genes. These characters may not be expressed uniformly, as their expression may be related to environmental or physiological factors. In particular, only a relatively small proportion of genes present in the genome are actually transcribed and translated into proteins at any given time, and environmental conditions, physiological age and the overall state of the plant can affect the factors that control the expression of such proteins. Thus both plant morphology and patterns of expressed proteins of genetically identical plants may be different if the plants are grown in different environments, or if they are analysed at different stages of growth. This makes comparisons based on these characters rather difficult, especially when data from different geographical areas are compared.

These limitations have led to investigation of methods of analysis based on the genome itself. As a first approach, the chromosomes can be examined using cytological techniques. These involve fixing the chromosomes during mitosis or meiosis, and then either staining them completely, or using stains for specific areas, such as Giemsa or C-banding. This approach provides data on both chromosome number and on structural abnormalities such as dicentric chromosomes, inversions and translocations. Significant alterations, in both chromosome number and structure, have been demonstrated as a result of regenerating plants from culture via a callus phase (Karp and Bright, 1985).

More recently, methods to analyse the genome at the DNA level have been developed. The advantages of using methods specifically to analyse the DNA are two-fold. First, and most importantly, the resolution that can be obtained is far greater than by any other method. Potentially, single base alterations can be identified, whether or not they occur within sequences that code for genes. Second, the DNA sequence is essentially the same in all cells in all tissues of the plant, and DNA can be extracted and stored for long

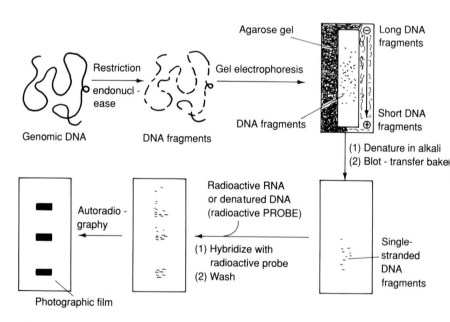

Figure 4.1 Representation of the procedures required for DNA molecular fingerprinting. DNA is extracted and cut into fragments by restriction endonucleases. The DNA fragments are separated by size using electrophoresis, denatured, transferred to membranes, and hybridized to radioactive probes. The bound radioactive probes are visualized by autoradiography.

periods of time by freezing. This represents a significant improvement over other methods, where tissue source and physiological age must be standardized for comparative purposes.

The molecular methods used are often referred to as 'DNA fingerprinting', and have many other applications apart from the assessment of stability for *in vitro* storage (e.g. Helentjaris *et al.*, 1985; Erlich *et al.*, 1986). The overall scheme that is followed is outlined in Figure 4.1. Molecular analysis relies on the specificity of enzymes called restriction endonucleases, which recognize specific base sequences ('recognition sites') at which they bind and cut the DNA (Figure 4.2). This produces fragments of DNA of discrete lengths that depend on the DNA sequence present. The fragments can then be separated by gel electrophoresis and analysed by visualization with specifically labelled DNA 'probes' (Figure 4.2). The more correct title for the approach is 'restriction fragment length' analysis. When the DNA fragments have been separated, on the basis of size, differences in DNA sequences will show up as fragments of different lengths, and will reflect events such as insertions, deletions, inversions and base alterations (if they occur at a restriction enzyme recognition site). Analysis of fragments can also be quantitative, to reveal amplification and reduction in the copy number of genes. The methylation pattern of certain bases can also be examined using isoschizomers of restriction enzymes – two enzymes which recognize the same site but which are affected differently by the methylation of bases at that site (Brown and Lörz, 1985). The methylation of DNA has been implicated in some studies on somaclonal variation and may be of interest to those studying stability in stored tissue. Although methylation is an epigenetic change, under certain circumstances it appears to be stable.

Whilst the methods for examining stability that involve analysis at the morphological, protein, and chromosome levels are useful, and provide a significant approach to assessing stability and plant screening, the undoubted advantages of restriction fragment length analysis will lead to increasing use of this new approach in the future. Because of this, and because many workers in this field may be unfamiliar with such techniques, fingerprinting methods are described in more detail.

4.3 USE OF DNA FINGERPRINTING TO ASSESS STABILITY

The only approach that can be used to be absolutely certain that a genome shows no variation is to sequence it entirely. This has not yet been achieved for any but the simplest organisms. From a practical viewpoint, DNA fingerprinting can provide a sufficiently high resolution analysis of genome organization, but it will not be complete. A further consequence is that,

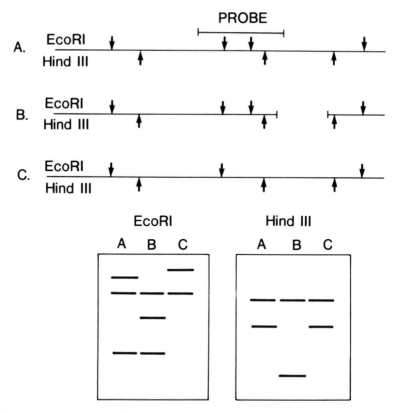

Figure 4.2 Diagram to illustrate how alterations in hybridization patterns can be produced. (A) shows the distribution of recognition sites (arrowed) for the restriction enzymes EcoR1 (above the line) and HindIII (below the line) and the pattern of bands visualized by the probe. (B) shows the effect of a deletion of a large fragment of DNA which alters the hybridization pattern produced by both enzymes. (C) shows the effect of a small alteration – possibly a single base change – which removes an EcoR1 restriction site, producing an altered hybridization pattern. In this case the HindIII restriction pattern is unaffected.

because of the relative complexity of the analysis, fingerprinting will not replace field selection for seed stocks. At present it is barely feasible to use protein or chromosome analysis routinely to screen the thousands of plants that may be produced for field selection. However, the DNA fingerprinting technique does appear to offer a unique opportunity to compare and contrast the effects of different tissue-culture regimes in terms of their effects on the genome. It is also applicable to the study of *in vitro* conservation methods, since these treatments can alter the physiology of tissues to the extent that effective comparison by phenotypic methods is not possible until a period of readjustment has occurred after transfer to normal growth conditions.

It is particularly important that the genetic fidelity of stock mother shoot cultures can be assessed before they are used as a source for micropropagation and dissemination. To assess stability after micropropagation, from a practical point of view every shoot culture propagated or stored need not be examined, but the number of plants analysed must be statistically significant, such that general conclusions on stability can be drawn. Also, as has been pointed out, a complete analysis is almost impossible to carry out, and so a decision must be taken on the number of combinations of restriction enzymes and DNA probes to use to obtain sufficient resolution of analysis. To help this decision, it is useful to study plants obtained from tissue-culture systems known to cause variation, both to show that the analysis is effective and to give an idea of the extent of variation that may be present, so that sufficient samples are taken for the study.

4.4 AN APPROACH TO DNA FINGERPRINTING

It is necessary to appreciate the manipulations required for DNA finger-printing, so these will be discussed in more detail in this section. As outlined in Figure 4.1, the various procedures involved are discussed under the following headings:

1. DNA isolation/purification.
2. DNA restriction, electrophoresis, blotting.
3. Probe isolation and labelling.
4. Hybridization.

These aspects will be considered in relation to analysis of plant genetic stability in *in vitro* conservation systems. Much of the theory and the basic techniques in molecular biology can be found in the practical manual of Maniatis *et al.* (1982).

4.4.1 DNA Isolation/purification

Methods for the isolation of plant DNA differ from those for other DNA sources because the plant cell wall provides an additional obstacle to extraction. This may be overcome by grinding the tissue in liquid nitrogen, to cause mechanical wall breakdown, before addition of detergents, proteinases and usually EDTA – a chelator of divalent cations (which are required by enzymes that degrade DNA). In standard methods, after extraction with phenol and chloroform to remove the protein contaminants, the DNA is usually purified by centrifugation in caesium chloride solution, where the DNA forms a discrete band because of its particular buoyant density. After dialysis to remove the caesium chloride, and precipitation of the DNA by addition of ethanol, an extremely pure sample of DNA can be obtained.

However, this method is time-consuming and expensive, and, where many DNA samples have to be prepared rapidly, various quicker extraction methods have been devised. Much time can be saved by eliminating the caesium chloride gradient centrifugation. When this is done, although DNA of such high purity may not be obtained, yields of DNA can be much higher.

For the purposes of restriction analysis, the DNA needs only to be pure enough to allow restriction enzymes to work. Thus such rapid extraction methods are useful for this work. In our laboratory we have applied a rapid 'miniscale' DNA extraction method, derived from a method used for maize restriction mapping (Dellaporta *et al.*, 1983) for the extraction of DNA from potato leaves, tuber sprouts and whole-shoot cultures. With slight modifications, this method has proved to be simple and rapid, yet produces DNA of good purity and quality.

4.4.2 DNA Restriction, electrophoresis, blotting

There are three different types of restriction endonuclease enzymes. The type usually applied in restriction analysis are 'type II' enzymes, which recognize specific four, six or eight base pair sequences in DNA, and cleave double-stranded DNA within that sequence. These enzymes exhibit very high specificity for their target sequence (recognition site) and a great many have now been characterized and are commercially available.

The choice of which enzymes to use is often made on the basis of cost and availability, since the distribution of recognition sites is essentially random within the genome. However, the choice between using enzymes recognizing four, six or eight base pair sites can be made depending on the resolution required, and, as the analysis of the fragments produced must be considered, the electrophoresis facilities available must also be taken into account. The greatest resolution is obtained by using 'four-cutters' (enzymes recognizing a four base pair sequence) because there are many such sites in the genome. The fragments so produced will be relatively small, and so small insertions and deletions can be resolved, and there is more chance of identifying single base alterations because more of the genome is present as recognition sites.

Conversely, use of an enzyme that recognizes an eight base pair sequence will require fewer probes, because large fragments of the genome are analysed at one time. As a result of this, only large alterations of the DNA will be visualized and complex electrophoresis systems must be used to resolve the fragments into discrete bands. As a compromise, 'six-cutters' are most often used for restriction fragment analysis. As well as being most readily available (and therefore cheaper), they usually produce fragments in the size range of 20 000 to 200 base pairs, which can be separated conveniently on agarose gels.

After digestion by restriction enzymes, the DNA is present as a mixture of linear double-stranded molecules of various lengths. These are most effectively separated by electrophoresis through agarose or polyacrylamide gels, in buffers near neutral pH, to maintain the negative charge on the DNA. The choice between agarose and polyacrylamide is based on the size of the DNA molecules to be separated, as the two systems have different properties.

Polyacrylamide gels can be made to resolve single base differences for sequencing uses, but, because they are difficult to handle below a concentration of 4% acrylamide, they do not efficiently resolve fragments above 1.5–2 kilobases. Also, the acrylamide monomer is toxic, so handling of the gels and their casting requires more care.

Conversely, agarose gels are much easier to prepare and handle. Agarose is a sulphated polysaccharide which only needs to be boiled in the buffer for a few minutes before being poured into slabs. The gels can be handled at agarose concentrations down to 0.5% or 0.6%, so fragments between 200 base pairs and 50 kilobase pairs can be separated using standard electrophoresis conditions. Even larger fragments can be resolved using a special system of applied voltages called 'reversed field' or 'pulsed' gel electrophoresis that enables separations of molecules of millions of base pairs in length.

The choice of gel system to be used is usually taken in conjunction with the restriction enzymes chosen. Four-cutters produce fragments too small to be resolved by agarose gels, and so polyacrylamide gels are required. Conversely, polyacrylamide gels cannot normally be used to resolve the fragments produced by six-cutters, and so agarose gels must be used. These considerations have led to most workers in the field using restriction analysis to use six-cutters, as agarose gels are much easier to handle.

Electrophoresis is usually carried out in Tris buffers at voltages below 5 V/cm. After electrophoresis, the DNA is visualized by staining in a 0.5 µg/ml ethidium bromide solution. This compound intercalates into the DNA helix and fluoresces when illuminated by short-wavelength ultraviolet light.

When genomic DNA is cut with restriction enzymes and separated by a gel, a smear is seen because the DNA has been cut into a range of sizes. As nothing can be seen in this smear, the DNA must be transferred onto a solid support and probed with a fragment of a labelled DNA to visualize specific sequences. This technique was developed by Southern (1975) and is referred to as 'Southern blotting' or 'Southern hybridization'. The basis for this is the transfer of the DNA from the gel to a solid support ('membranes' or 'filters'), thus preserving the position of the fragments as they were in the gel, yet enabling hybridization reactions to be performed. The transfer of DNA onto supporting membranes was originally done by capillary action – a stack of blotting paper placed on top of the gel – filter sandwich pulled buffer through the gel and the filter, bringing the DNA with it from the gel and leaving it bound to the filter. It is now routinely done by electric current (electroblotting) or by vacuum transfer. The original material used as the filter was nitrocellulose, but this was fragile and nylon membranes are now used to enable repeated cycles of hybridization to be carried out. Because the same probe can give different hybridization patterns when used to probe DNA cut with different enzymes, a number of filters are prepared using DNA from the same source but cut with different restriction enzymes. Since the filters can be re-probed, it is more convenient to use many probes and fewer restriction enzymes. However, cutting of DNA by more than one enzyme is necessary for adequate resolution of analysis.

After blotting, the DNA is bound irreversibly to the filter by baking at 80°C or by ultraviolet irradiation. The dry filters can then be stored indefinitely until required.

4.4.3 Probe isolation and labelling

Probes used to identify specific DNA fragments by hybridization are of two types – genomic clones (fragments of nuclear DNA) and cDNA clones (DNA copies of mRNA molecules). Both genomic and cDNA clones can be maintained in plasmid or phage vectors. Libraries of both types of probe can be made relatively simply, and many commercial kits are available. The two methods produce different types of probes which may be used for different purposes. Genomic libraries will contain many repetitive probes because repetitive sequences constitute the largest proportion of plant genomic DNA. Such probes will hybridize to many fragments on the filters and produce very complex patterns – so complex that it may be difficult to resolve single bands and almost impossible to detect subtle changes in the genome. cDNA libraries will contain predominantly unique or low copy number sequences representing expressed genes and so will show fewer bands on the filter. Although use of cDNA probes will help the identification of small changes, the proportion of the genome 'covered' by each probe will be relatively small and many more probes must be used.

By screening genomic libraries for middle and low copy number sequences, the hybridization patterns obtained can be simplified (Landsmann and Uhrig, 1985). Similarly, cDNA clones from sequences from multigene families can increase the proportion of the genome covered by these probes. In this way, careful selection of the probes can reduce the time and effort required for analysis.

Probes can be isolated by various methods, which depend mostly on the vector used for cloning. Only small amounts of probe DNA are required for labelling and hybridization, so rapid miniscale procedures are convenient (Birnboim and Doily, 1979).

The labelling of probes involves synthesis of DNA to incorporate one or more nucleotides which are labelled with a reporter group, usually a radioactive isotope. Originally this was achieved by the method of 'nick-translation' (Rigby *et al.*, 1977), where DNA polymerase I was used to synthesize a new strand of DNA from nicks produced by addition of a small amount of a DNase enzyme. A requirement for these two enzymes was that the DNA should be very pure, and thus caesium chloride purification was usually required. A more widely used modification of this method (Feinberg and Vogelstein, 1983) uses only the large fragment of DNA polymerase I (Klenow fragment) to synthesize labelled strands of DNA using random hexanucleotides as primers ('oligolabelling'). This method does not require DNA of such high purity, because the Klenow fragment is not as sensitive as the holoenzyme. A consequence of this is that not only can the rapid plasmid isolation protocols be used for probe isolation, but also fragments of plasmids can be labelled in the presence of agarose contaminants.

To reduce background due to non-specific hybridization, it is often preferable to separate cloned genomic or cDNA fragments from the vector before labelling rather than labelling the whole plasmid. Isolation of cloned fragments is quite straightforward when using oligolabelling. The plasmid is cut with an appropriate restriction enzyme, and then the cloned fragment is separated from the carrier plasmid on a gel. After staining the gel, the segment containing the desired fragment is cut out and boiled in a small volume of distilled water, and can then be stored at $-20°C$. When required, the sample is thawed, and an aliquot is removed and boiled before adding to the labelling reaction directly. Oligolabelling can also produce DNA with a greater specific activity of label than that obtained by nick translation. This allows detection of the bands in a shorter time, as well as improving the resolution by giving a better signal-to-noise ratio.

As DNA hybridization methodology has been applied in industry (van Brunt and Klausner, 1987), alternative labels and detection systems used have been examined. However, ^{32}P is usually chosen as the label for probes, because the hybridized bands are visualized by autoradiography. Non-radioactive reporter molecules, mostly based on biotin, can be visualized by staining. In the latter case, since the stain cannot be washed off (as can the hybridizing DNA), the filters can only be used once. In contrast, using radioisotope-labelled probes, the filters can be re-probed via several cycles of hybridization. This shortens the work required for most restriction fragment analyses.

4.4.4 Hybridization

Hybridization is the process by which the labelled probe binds to complementary DNA on the filter, enabling visualization of specific DNA fragments. The binding of the probe to the target sequence utilizes the same hydrogen bonding as in duplex DNA molecules, and is thus very specific (Hames and Higgins, 1985). The target DNA is denatured in the gel before transfer to the filter, and, similarly, the probe DNA is denatured before being applied to the filter in solution. This probe solution is left in contact with the filter for long enough to allow hybridization of the probe to target sequences. In addition, non-specific hybridization occurs between similar, but not identical, sequences. This must then be washed off under more 'stringent' conditions than those used for initial hybridization. The stringency is increased by raising the temperature or lowering the salt concentration of the wash buffer. By careful control of the temperature and ionic strength of the buffer, discrete amounts of 'mismatch' can be allowed, enabling hybridization to proceed with probes from different organisms ('heterologous probes'). Often it is possible to estimate the degree of mismatch by sequential washes of progressively higher stringency until the probe is washed off. Some factors that affect this process are considered below.

Before use, the dry filter is first pre-hybridized in hybridization solution without the probe. This process consists of adding fairly high salt concentrations in phosphate buffer, detergent, Denhardts solution (a mixture of ficoll, polyvinylpyrrolidone and bovine serum albumen) and non-homologous random-sequence DNA, usually from salmon or herring sperm. The random-sequence DNA blocks binding of repetitive sequences to which the probe might otherwise bind and cause background signals. Other compounds can be added to hybridization solutions, such as formamide (to reduce the temperature required for the hybridization) and dextran sulphate (to reduce the effective probe solution volume).

After pre-hybridization, the probe is added. The time required for hybridization depends on the amount of probe, its complexity (usually simplified to its length) and the volume of the hybridization solution. To reduce the volume of liquid involved, hybridization is usually performed in sealed plastic bags. The main requirement is that a film of solution containing the radioactive probe covers the whole surface of the filter, to ensure even hybridization.

After hybridization, the probe is removed and may be re-used. Filters are washed in salt solutions with detergent. For hybridization of homologous probes to blots of genomic DNA, the washing stringency can be taken to high levels to ensure a low background, e.g. low salt concentrations and quite high temperatures (65°C) will allow less than 5% mismatch between probe and target sequence.

The bound probe then needs to be detected by autoradiography. The filter is wrapped in thin plastic film ('clingfilm') and placed in a cassette in contact with X-ray-sensitive film. After 1–2 days exposure, the film is developed and the position of the labelled probe is revealed as blackened areas of film. ^{32}P, the commonest isotope used, emits particles of such high energy that they can pass through the film without affecting it. This problem is corrected by using 'intensifying screens' above and below the filter–film sandwich which absorb the particles and emit many slower particles, which result in image formation on the film. In addition, the autoradiography cassette is usually placed in a deep freeze at −75°C, to slow the particles. After autoradiography, the probe can be easily washed off the filter in hot formamide or low ionic strength buffers and new probes can then be re-hybridized to the filter. As long as they are not allowed to dry out, nylon filters can be re-hybridized ten times or more in this way, with no loss of resolution.

4.5 AN INVESTIGATION OF GENETIC STABILITY OF POTATO AFTER MICROPROPAGATION OR STORAGE *IN VITRO*

The techniques described in the previous sections are being used to analyse DNA extracted from potato plants that have undergone micropropagation or storage (at reduced temperature or under osmotic stress). The types of tissue-culture treatment that have been used are shown in Figure 4.3. In

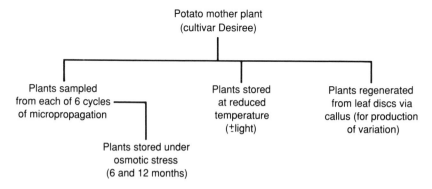

Figure 4.3 Tissue-culture regimes used to generate plants after micropropagation or storage, or to produce variants via a callus phase.

83

this work, using the potato cultivar Desiree, from one 'mother' plant a range of plants, as illustrated in Figure 4.3, have been generated. Six lines have been propagated, by serial subculture of nodes, through six cycles of subculture. Whole plants have been generated at each stage, and the DNA extracted. Similarly, DNA has been extracted from plants grown from cultures stored at 4–8°C, and under osmotic stress (with 4% mannitol) to induce slow growth. In addition, to provide control tissue that will show variation, plants have been regenerated from leaf discs of the mother plant via a callus phase (Wheeler *et al.*, 1985). Other known somaclonal variants have also been examined. This material includes protoplast-derived plants produced at Rothamsted (Foulger, 1987), and somaclonal variants, re-generated from leaf discs, that were identified as variants by breeders in field experiments (Evans *et al.*, 1986).

The analytical scheme that is being applied to these plants is outlined in Figure 4.4. In most cases, the tissues used for DNA extraction have been

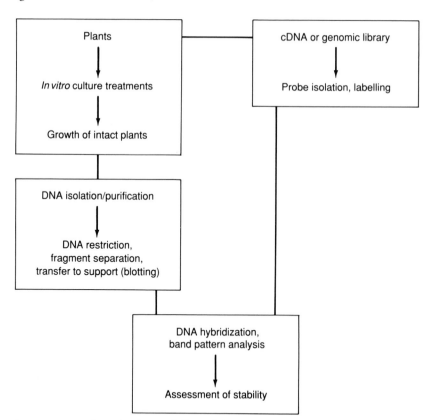

Figure 4.4 Analytical scheme followed for assessment of variation/stability by DNA fingerprinting.

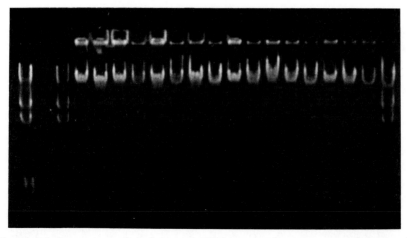

Figure 4.5 Appearance of total genomic DNA extracted, run on an acrylamide gel, and stained with ethidium bromide. High molecular weight DNA is present. (Markers: 0.5 µg of HindIII restricted lambda DNA).

leaves from pot-grown plants, but, in some cases, shoots propagated in liquid media as shaker cultures have been used. The latter approach provides adequate tissue for extraction soon after the experimental treatment (3–4 weeks), and has the advantage that further growth can be induced to produce *in vitro* tubers for protein analysis and storage (Tovar *et al.*, 1985).

The appearance of total extracted DNA, run on a gel and stained with ethidium bromide, is shown in Figure 4.5. This step is normally carried out to check DNA purity. The appearance in gels of DNA that has been cut into fragments by restriction endonucleases is shown in Figure 4.6. Such DNA, after transfer to a membrane, hybridization with a radioactive (^{32}P) probe, and autoradiography, is shown in Figure 4.7. Three marker tracks with different probe copy numbers ('reconstructions') are also included. It shows that 11 different cultivars can be readily identified by their different patterns of bands using just one combination of probe and restriction enzyme.

When somaclonal variants of one original genotype are analysed, it is also possible to identify changes in the patterns of bands. This is illustrated in Figure 4.8, in which the loss of one band in a somaclonal variant of the cultivar Desiree is evident. Another change which is frequently encountered with variant material is in band intensity, as shown in Figure 4.9. This indicates an alteration in the copy number of the gene involved, i.e. amplification or deletion of copies. This finding agrees with those of other groups that have studied somaclonal variation (e.g. Landsmann and Uhrig,

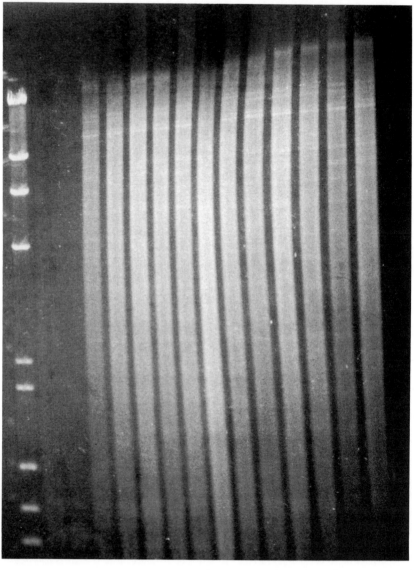

Figure 4.6 Appearance of genomic DNA after cutting into fragments by a restriction endonuclease, gel separation, and ethidium bromide staining. The DNA fragments are present as a continuous range of sizes. (Molecular size markers are HindIII restricted and lambda DNA, and HaeII restricted φ × 174 DNA.)

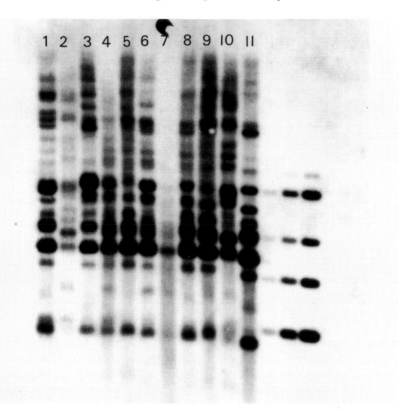

Figure 4.7 Banding patterns obtained after probing restricted genomic DNA of 11 cultivars of potato with ^{32}P labelled probe of patatin, the major tuber protein. Lanes: 1, Vanessa; 2, Redskin; 3, Ulster chieftain; 4, Eigenheimer; 5, Rodestar; 6, Furore; 7, Duke of York; 8, Blanka; 9, Romano; 10, Draga; 11, Desiree; and three marker lanes.

1985). In our work, most variation has been identified in plants from shoots regenerated directly from leaf-disc callus. Variants selected in the field, which have undergone some 'selection' for normal growth by breeders, exhibit less variation in bands.

The overall results of restriction fragment analysis carried out on over 200 plants from the various *in vitro* regimes outlined in Figure 4.3 has confirmed that plants studied after micropropagation are more stable genetically than plants regenerated via a callus phase. In fact, no variant plants were identified out of a total of 167 plants regenerated from organized meristems, whereas 6 variant plants were identified from a population of 46 regenerated from leaf disc callus (Potter, 1989; Potter and Jones, 1990). This is clearly not conclusive evidence that all plants derived

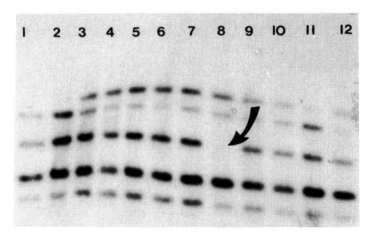

Figure 4.8 Loss of one band arrow in the pattern of fragments of a somaclonal variant of cultivar Desiree probed with a random cDNA clone.

from micropropagation are stable, but it does show the value of the restriction fragment analysis approach, in that significant differences between the populations were found even with this small sample number.

Other aspects that may have bearing on apparent or real variation from culture, and should be considered, include possible genotypic components of stability (e.g. Jones and Karp, 1985), breakdown of chimaeral varieties to yield plants from one or other cell type, and the possibility of mislabelling tubes. A number of apparent cases of variation can be tracked down to the latter cause!

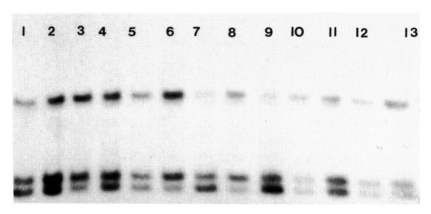

Figure 4.9 Changes in band intensities of a somaclonal variant of cultivar Desiree probed with a genomic clone for rRNA coding sequences.

Acknowledgement

This report emphasizes that it is important that the stability of stock mother shoots, maintained *in vitro*, should be assessed before the culture is used for bulking up plants for commercial use. The normal methods of assessing uniformity and stability usually require field growth of plants. This approach is ruled out for crops such as potato, because plants in the field would be liable to become infected, and also the process of assessment is time-consuming. We have established that an alternative approach, that of examining protein patterns of microtubers produced *in vitro*, is inadequate (unpublished observations), because the protein patterns of microtubers differ from those of field-grown plants. DNA fingerprinting can already be applied to check that a shoot culture is the correct cultivar (e.g. Figure 4.7). With suitable refinements, the same approach can also be used to look for any variation within a specific cultivar.

4.6 CONCLUSION

At present there is little information available on the stability of plant material following micropropagation or storage in tissue culture. Restriction fragment analysis offers the potential to examine stability at the DNA level, but techniques are still being developed, and the approach has not yet been applied in detail to examine stability following tissue-culture propagation and storage treatments. The results outlined here indicate that the approach promises to provide the most detailed method of genome analysis, so that slight genetic differences between genotypes can be identified.

ACKNOWLEDGEMENT

We thank the International Potato Centre (CIP) for financial support, and Dr W. Park for the patatin probe.

REFERENCES

Benson, E. E. and Withers, L. A. (1987) Gas chromatographic analysis of volatile hydrocarbon production by cryopreserved plant tissue cultures: a non-destructive method for assessing stability. *Cryo-Letters*, **8**, 35–46.

Birnboim, H. C. and Doly, J. (1979) A rapid alkaline extraction procedure for screening recombinant plasmid DNA. *Nucleic Acids Res.*, **7**, 1513.

Brown, P. T. H. and Lorz, H. (1985) in *Somaclonal Variation and Crop Improvement* (ed. J. Semal), Martinus Nijhoff, Amsterdam, pp. 148–159.

Dellaporta, S. L., Wood, J. and Hicks, J. P. (1983) in *Plant Molecular Biology – a Laboratory Manual*, Cold Spring Harbour Laboratory, New York, pp. 36–37.

Denton, I. R., Westcott, R. J. and Ford-Lloyd, B. V. (1977) Phenotypic variation of *Solanum tuberosum* L. cv. Dr. McIntosh regenerated directly from shoot tip culture. *Potato Res.*, **20**, 131–136.

El-Gizawy, A. M. and Ford-Lloyd, B. V. (1987) An *in vitro* method for the conservation and storage of garlic germplasm. *Plant Cell Tissue and Organ Culture*, **9**, 147–150.

Erlich, H. A., Sheldon, E. L. and Horn, G. (1986) HLA typing using DNA probes. *Bio/Technology*, **4**, 975–981.

Evans, N. E., Foulger, D., Farrer, L. and Bright, S. W. J. (1986) Somaclonal variation in explant derived potato clones over 3 tuber generations. *Euphytica*, **35**, 353–361.

Feinberg, A. P. and Vogelstein, H. (1983) A technique for radiolabelling DNA restriction endonuclease fragments to high specific activity. *Anal. Biochem.*, **132**, 6–13.

Foulger, D. (1987) Protoplast regeneration and somatic hybridisation of potato (*Solanum tuberosum*), PhD thesis, Rothamsted/CNAA.

Fowler, M. W. (1984) *Biotechnol. Genet. Eng. Rev.*, **2**, 41–60.

Hames, B. D. and Higgins, S. J. (eds) (1985) *Nucleic Acid Hybridisation*, IRL Press, Oxford.

Helentjaris, T., King, G., Slocum, M., Siedenstrang, C. and Wegman, S. (1985) Restriction fragment polymorphisms as probes for plant diversity and their development as tools for applied plant breeding. *Plant Mol. Biol.*, **5**, 109–118.

Hussey, G. and Stacey, N. J. (1981) *In vitro* propagation of potato (*Solanum tuberosum* L.). *Ann. Bot.*, **48**, 787–796.

Jones, M. G. K. and Karp, A. (1985) Plant tissue culture technology and crop improvement. *Adv. Biotechnol. Processes*, **5**, 91–121.

Karp, A. and Bright, S. W. J. (1985) *On the Causes and Origins of Somaclonal Variation* (*Oxford Surveys of Plant Molecular and Cell Biology*, Vol. 2), Oxford University Press, pp. 199–234.

Landsmann, J. and Uhrig, H. (1985) Somaclonal variation in *Solanum tuberosum* detected at the molecular level. *Theor. Appl. Genet.* **71**, 500–505.

Larkin, P. and Scowcroft, W. R. (1981) Somaclonal variation – a novel source of variability from cell cultures for plant improvement. *Theor. Appl. Genet.*, **60**, 197–214.

Maniatis, T., Fritsch, E. P. and Sambrook, J. (1982) *Molecular Cloning*. Cold Spring Harbour Laboratory, New York.

National Institute of Agricultural Botany (1975) *Guide to the Identification of Potato Varieties*, NIAB, Cambridge, UK.

References

Potter, R. H. (1989) Genetic stability of the potato (*Solanum tuberosum* L.) after *in vitro* culture. PhD thesis, University of London.

Potter, R. H. and Jones M. G. K. (1990) Analysis of genetic stability of potato *in vitro* by molecular and phenotypic methods. (Submitted.)

Rigby, P. W., Dieckmann, J. M., Thodes, C. and Berg, P. (1977) Labelling deoxyribonucleic acid to high specific activity *in vitro* by nick translation with DNA polymerase J. *J. Mol. Biol.*, **113**, 237–251.

Southern, E. M. (1975) Detection of specific sequences among DNA fragments separated by gel electrophoresis. *J. Mol. Biol.*, **98**, 503–517.

Stegeman, H. and Schnick, S. (1985) *Index 1985 of European Potato Varieties*, Paul Parey, Berlin.

Tovar, P., Estrada, R., Schilde-Rentschler, L. and Dodds, J. H. (1985) Induction and use of *in vitro* tubers. *CIP Circular*, **13**(4), 1–6.

Van Brunt, J. and Klausner, A. (1987) Pushing probes to market. *Bio/Technology*, **5**, 211–221.

Wheeler, V. A., Evans, N. E., Foulger, D., Webb, K. J., Karp, A., Franklin, J. and Bright, S. W. J. (1985) Shoot formation from explant cultures of 14 potato cultivars and studies of the cytology and morphology of regenerated plants. *Ann. Bot.*, **55**, 309–320.

Wright, N. S. (1983) Uniformity among virus-free clones of ten potato cultivars. *Am. Potato J.*, **60**, 381–388.

5

Potato germplasm conservation

**J. H. DODDS, Z. HUAMAN
and R. LIZARRAGA**

5.1 INTRODUCTION

The last decade has seen an ever-growing awareness of the value of germplasm conservation. As new varieties comprise an increasingly large proportion of cultivated crops, the genetic diversity of many major crops is being eroded. A number of initiatives have been taken by international organizations to collect, evaluate and preserve wild and primitive geno-types, in order to preserve these gene pools containing the breeding materials of the future (Frankel and Hawkes, 1975; Nitzsche, 1983; Wilkins and Dodds, 1983).

Many major crop plants are propagated routinely by seed, i.e. rice, maize and wheat; however, a number of crops have to be propagated asexually because either the plants do not produce seeds or the crop depends on the performance of a selected genotype. The preservation of the latter type of germplasm requires the material to be maintained in a clonal form by an appropriate means of vegetative propagation.

Although in the potato it is possible to produce botanical seed (true seed), this seed is highly heterozygous. True seed is an effective way of conserving the total gene pool; however, selected individual genotypes have to be conserved as a clonal collection (Morel, 1975).

The clonal collection of potatoes maintained in the field at the International Potato Center (CIP) numbers over 3000 accessions. Annually, the collection has to be planted, maintained and harvested, and the tubers stored until the following season. A number of problems may be encoun-tered with field maintenance of this type of collection, and these include:

1. Cost of labour and space.
2. Risk of infection by bacteria, fungi and viruses.
3. Risk of damage by insect pests.

In Vitro *Methods for Conservation of Plant Genetic Resources. Edited by John H. Dodds.*
Published in 1991 by Chapman and Hall, London.   ISBN 0 412 33870 X

4. Risk of damage by hail, frost, wind, etc.
5. Seasonal availability of materials.

Some of these problems have added to the need to devise a method for clonally conserving potato germplasm *in vitro*.

Over the last ten years a great deal of research has been carried out on the *in vitro* culture of potato, a significant part of which focused on *in vitro* germplasm maintenance (Henshaw and O'Hara, 1983; Dodds, 1986; Nitzsche, 1983; Withers, 1983). A wide range of methods is available, each with its own advantages and pitfalls. This chapter will review the methods available and their comparative advantages, and will describe a standard method now adopted at CIP for the *in vitro* conservation of its large germplasm collection.

For many years *in vitro* methods have been used for conservation of a wide range of plant species. However, it is only in recent years that attention has been given to the use of *in vitro* methods for germplasm collection.

Problems often exist in moving vegetative material after a collecting trip, either for quarantine reasons, or sometimes on the grounds of bulk of material, e.g. coconuts. The International Board of Plant Genetic Resources (IBPGR) has been active in encouraging research in this area. The IBPGR (1984) published a report on the potential use of this technique. The simple methods applied so far depend on surface sterilization with fungicide/bactericide followed by washing in purified water. Single node sections are then inoculated onto tubes of pre-prepared media that contain antibiotics and fungicides. This method has been used and published for cocoa by Yidana *et al.* (1985). The CIP (International Potato Center) has also been active in developing this technique for the collection of sweet potato germplasm.

5.2 METHODS OF *IN VITRO* GERMPLASM STORAGE

5.2.1 Use of growth regulators

Once material has been introduced to *in vitro* culture, the objective is to restrict the amount of labour required to keep the material in this state. If no suitable growth-restriction method is found, then the amount of labour required for subculturing would be enormous. *In vitro* potato cultures on a standard propagation medium (MS salts + source) under optimum growth conditions require subculture at least once every two months (Roca *et al.*, 1978).

A range of chemical growth retardants has been tested. The logic of using these compounds is simply to reduce the overall growth rate of the *in vitro* plants and lengthen the time of subculturing. Maleic hydrazide (MH) is the active compound in a number of commercially available growth retardants and has been shown to promote tuberization in cultured stem sections of *S. tuberosum* cv. British Queen (Harmey *et al.*, 1966). Diaminozide (B995) is used extensively as a foliar spray on ornamental plants such as chrysanthemums and azaleas. Humphries and Dyson (1967) reported a 10% increase in tuberization after spraying it on plants of *S. tuberosum* cv. Majestic. Also, C_6–C_3 phenolic compounds such as *trans*-cinnamic acid (TCA) have been shown to stimulate tuberization from stem cuttings *in vitro* (Paupardin and Tizio, 1970). Abscisic acid (ABA), which is present in potato tubers, is involved in the control of dormancy and generally acts as a natural growth retardant (Addicott and Lyon, 1969). In our laboratory, preliminary results have shown that the interval between transfers could be extended by adding ABA to the culture medium (Westcott *et al.*, 1977a). Sadatz and Standke (1978) reported that the addition of *N*-dimethyl succinamic acid at concentrations up to 100 mg/l allowed nodal cultures of potato to survive storage at 2–7°C for 12 months and subsequently to be transferred to soil. Chlorocholine chloride (Phosphon), a commercial stunting agent frequently added to compost to dwarf ornamentals, has also been investigated (Stecco and Tizio, 1982). An alternative approach to retarding growth was to increase the osmotic pressure of the medium, by adding the metabolically inactive sugar alcohol mannitol, in order to reduce the water available to growing cultures.

Table 5.1 shows the effect of the addition of ABA on the survival of nodal cultures from five cultivars. ABA enhanced survival at all concentrations, but the lower levels of 5 mg/l and 10 mg/l were most successful. In the presence of ABA, the plantlets became stunted with reduced internodes, some were chlorotic and root development was retarded compared to controls. There were no apparent differences in survival between cultivars, but the degrees of stunting and chlorosis did vary.

Table 5.2 shows the results of a similar experiment, the effect of other growth retardants on the survival of nodal cultivars. The results are not

Table 5.1 Effect of the addition of abscisic acid (ABA) on mean percentage survival of single nodal cultures from five potato cultivars at 22°C*

Storage period (months)	ABA concentration (mg/l)				
	0	5	10	20	50
6	52	96	88	96	68
12	17	66	61	34	27

* 3.5 ml MS medium containing no hormones other than ABA, 16 h photoperiod 1000 lux ($24\ W\ m^{-2}$). Total cultures 250 of 3.5 ml. Courtesy Westcott, R. J. (1977).

Table 5.2 Effect of the addition of four growth retardants on the mean percentage survival of single nodal cultures from five potato cultivars after six months at 22°C[a]

Retardant concentration (mg/l)	Retardant			
	B995[b]	TCA[c]	Phosphon D	MH[d]
0.0	45			
0.1	35	50	65	45
1.0	40	50	85	40
3.0	40	40	85	40
10.0	40	50	40	60
50.0	60	40	–	35

[a] 3.5 ml MS medium containing no hormones, 16 h photoperiod 1000 lux. Total cultures 200 of 3.5 ml MS.
[b] Dimethylamino succinamide.
[c] Trans-cinnamic acid.
[d] Maleic hydrazide.
Courtesy Westcott, R. J. (1977).

unequivocal but 50.0 mg/l B995, 1.0 mg/l and 3.0 mg/l Phosphon D and 10.0 mg/l MH allowed increased survival compared to controls. The morphological effects of phosphon D and MH were similar to those of ABA, i.e. stunting and poor root development. TCA and B995 did not produce changes in the growth habit of the cultures.

There are several disadvantages to using growth retardants *in vitro* as a method of germplasm conservation, as follows.

(a) Physiological
If the plants are grown for protracted periods of time in the presence of growth retardants, it is probable that pronounced physiological effects, such as tuberization, will be induced. In some cases these changes could be so pronounced that it might be difficult, if not impossible, to regenerate a physiologically normal plant at the end of the storage period.

(b) Selection
The application of a stress, in this case the application of a growth retardant, obviously allows the possibility for a type of *in vitro* selection (Maliga, 1978). A cell or small group of cells that are resistant or tolerant to the growth regulator would grow more rapidly. It is possible that, by a continuation of this process, one would eventually be selecting lines with resistance or tolerance to the growth retardant.

(c) Mutation
Some evidence exists, although it is not well substantiated, that some growth retardants have mutagenic properties. If the objective of clonal

conservation is to maintain a specific genotype, then obviously any system which induces mutations is unacceptable.

There are, however, advantages to the use of growth-retarding compounds. They represent a simple, efficient and relatively cheap way of modifying an existing growth medium to convert it to a conservation medium. The importance of germplasm stability is so overriding, however, that to date relatively few institutions have adopted the use of plant growth regulators as a routine way of maintaining an *in vitro* germplasm collection.

5.2.2 Minimal-growth media and restrictive growth conditions

The nutrition of plants *in vitro* has many parallels with the nutrition of plants *in vivo*. The needs for water, light and inorganic nutrition are universal. It is feasible, therefore, to modify the chemical composition of the culture medium in such a way as to limit growth, either by the absence or reduction of a specific chemical, or by addition of an extra modifying agent. Let us look at a few options that are available and the nutritional background to them.

Limiting the nutrient supply, i.e. nitrogen or magnesium, obviously leads to symptoms of nutrient deficiency and a corresponding reduction in growth rate. Care must be exercised, however, as too severe a stress will kill the plant.

Altering the available carbon source, either as a nutritional factor or as an osmotic factor, can have a very marked effect on growth rate. The use of different concentrations of sucrose has yielded interesting results. An alternative is the inclusion of a non-metabolizable sugar alcohol such as

Table 5.3 Effect of the addition of mannitol on the mean percentage survival of single nodal cultures from five potato cultivars at 22°C*

Storage period (months)	Mannitol concentration (%)				
	0	1	3	6	10
6	62	100	100	100	66
12	16	34	64	61	14
	Number of transferable nodes after 12 months				
PHU 5037	29	20	33	36	0
STN Pitiquiña	10	0	53	11	1
ADG Chata Blanca	17	62	71	19	4
TUB × ADG ariva	0	16	13	5	0
TUB Dr McIntosh	0	1	0	20	2

*3.5 ml MS medium containing no hormones, 16 h photoperiod 1000 lux (24 W m^{-2}). Total cultures 250. Courtesy Westcott, R. J. (1977).

sorbitol or mannitol in the medium. Table 5.3 shows the effect of mannitol on growth of *in vitro* shoot cultures of potato. This additive is a very effective way of regulating *in vitro* plantlet growth.

5.2.3 Growth regulation at reduced temperature

Plants, like most organisms, live within a fairly restricted temperature range. The growth processes are under the control of a large number of regulatory enzymes. Biochemically, each enzyme has a well-defined temperature optimum which affects the temperature optimum of the plant. It follows, therefore, that if the *in vitro* plants are maintained at temperatures either significantly above or below this optimum their growth will be restricted. Of course, care must be taken not to overstress plants. If temperatures fall much below 3°C, then frost damage will kill plants. Likewise, temperatures much above 28°C will cause excessive heat stress. Within this range of temperatures, plants will survive and grow; however, the total life expectancy changes markedly on either side of the 6°C mark.

Although some experiments were performed to study survival of material maintained below 0°C down to −12°C, the success of these experiments was very limited. However, more success has been achieved by dropping to extremely low temperatures (−196°C) using cryopreservation techniques (Bajaj, 1981; Kartha, 1985; Westcott *et al.*, 1977b; Kartha *et al.*, 1979).

5.3 TYPE OF PLANT MATERIAL TO BE CONSERVED

5.3.1 Cells and protoplasts

Although isolated cells and protoplasts are excellent materials for cryopreservation studies (Kanai and Edwards, 1973), there are major problems with the genetic stability of this material (Larkin and Scowcroft, 1981; Scowcroft and Larkin, 1982). Plants that are regenerated from conserved cells or protoplasts will pass through a callus stage during the regeneration process. It is known that in the case of potato the irregularities of cell division caused during callus growth result in both genotypic and phenotypic variation (Thomas, 1981). In the short term, therefore, it seems that these types of plant materials are unsuitable for conservation of potato genetic resources.

5.3.2 Callus

As indicated above, callus cultures of potato are known to be unstable and produce genetic variants. We cannot, therefore, see that use can currently be made of these cultures for conservation purposes. Care must also be exercised that there is no callus formation in other systems, i.e. meristem regeneration, or the same problem will be encountered.

5.3.3 Tuber slices

Although the use of tuber slices has been reported as a possible method of *in vitro* conservation of potato genetic resources (Silva, 1985), there are still problems with the genetic stability of this system. Although it may be possible to store this material, the regeneration phase normally involves the formation of a small callus followed by plantlet regeneration. If the objective is the storing of specific gene combinations, this type of variation is undesirable.

5.3.4 Isolated meristems

Within the limits of our present detection systems, meristem cultures are a source of regenerating genetically stable material so long as callus formation is avoided. Their size also allows some limited success in cryopreservation (Bajaj, 1981). However, the successes have been with a limited number of genotypes and the percentage success not much greater than 30%. It is not possible, therefore, to currently use this method as a routine way of storing large germplasm collections. However, this area still has great potential.

5.3.5 *In vitro* shoot cultures

At the present time this represents the most reproducible way of conserving a genetically stable germplasm collection. Slow growth, either by temperature reduction, growth inhibitors or minimal-growth media, can be applied to large germplasm collections.

5.3.6 *In vitro* tubers

Preliminary experiments are under way to study the effectiveness of *in vitro* tubers for germplasm conservation. Figure 5.1 shows *in vitro* tubers that have been on storage medium and maintained at 6°C for one year. It can be seen that this material has grown less than a comparable shoot culture

Figure 5.1 *In vitro* tuber after 12 months storage on conservation medium (containing mannitol) at 6°C in dim light (16 h day).

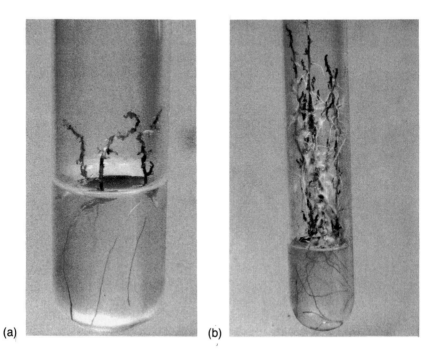

(a)   (b)

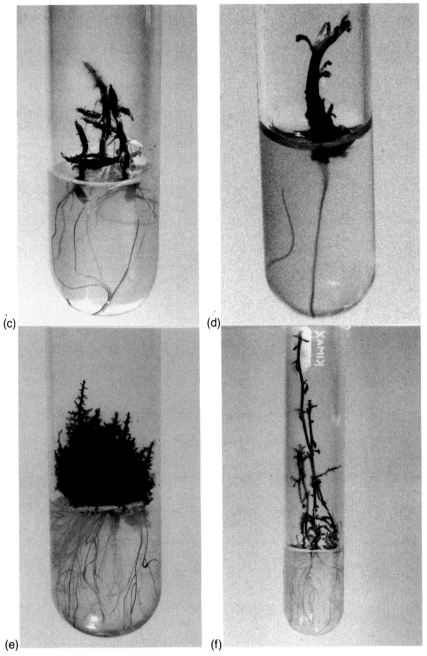

(c)   (d)

(e)   (f)

Figure 5.2 A range of different genotype responses. All materials in test tubes were stored at 6°C on conservation medium at 1000 lux (16 h day). The responses are (a) short internodes, (b) root hair formation, (c) tuberization, (d) browning and yellowing, (e) extensive branching, (f) elongated growth.

(Figure 5.2). This may offer an opportunity for extending the storage period to four or five years.

5.4 STANDARD INTERNATIONAL POTATO CENTER STORAGE TECHNIQUE

This technique involves using both a minimal-growth medium and a reduced incubation temperature to restrict growth. The medium composition is as shown in Table 5.4 and includes both sucrose and mannitol as an osmotic stress to inhibit growth. The cultures are maintained at $6°C \pm 1°C$ and 1000 lux, with a 16 h day.

Plant material from the field-maintained collection first has to be introduced to *in vitro* conditions. This can be done either from stem cuttings or tuber sprouts. The introduction process is shown in Figure 5.3, and involves the following steps:

1. Excision of sprouts/stem segments,
2. Surface sterilization,
3. Dissection of shoot tips,
4. Placement onto standard propagation medium,
5. Pregrowth at 20°C,
6. Transfer of nodal cutting to storage medium after maintaining for two weeks at 20°C,
7. Transfer to 6°C storage,
8. Continued observation of cultures to characterize growth and determine transfer date.

Using this medium and temperature, cultures can be maintained for at least 24 months and in many cases for longer between subcultures. There is,

Table 5.4 Murashige and Skoog (1962) medium was used with the following additions

Compound (mg/l)	Single-node propagation	Liquid-culture propagation	Conservation
Thiamine-HCl	0.4	0.4	0.4
Ca-Pantothenic acid	2.0	2.0	
Gibberellic acid	0.25	0.4	
Benzyl amino purine		0.5	
Naphthalene acetic acid	0.01		
Inositol	100.0	100.0	100.0
Sucrose (%)	3.0	2.0	3.0
Mannitol (%)			4.0
Agar (%)	0.8		0.8

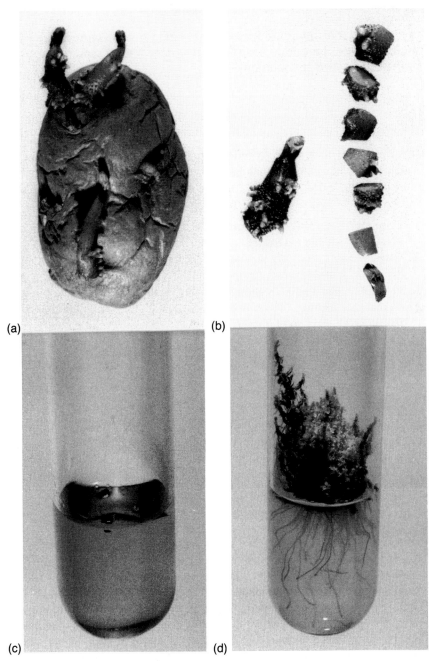

(a)

(b)

(c)

(d)

Figure 5.3 A photographic sequence of the stages involved in the introduction of a genotype to *in vitro* storage. These steps are (a) selection of material, (b) dissection of shoot tip (c), pregrowth on normal propagation medium, (d) transfer of nodes to conservation medium and transfer of tuber to 6°C storage.

however, a large difference in genotype response, and Figure 5.2 shows a number of genotypes that have been in storage for the same time period under similar storage conditions.

5.5 EVALUATION OF CONSERVED MATERIAL AND DATABASE

It is important that plant material is not simply transferred to a storage medium, put in a cold room at 6°C and forgotten. The plant material must be constantly monitored. At CIP we monitor the material visually at least once a month for the following characters.

1. Contamination: Any contaminated plantlets must be removed from the growth room and destroyed.
2. Growth rate: Analysis is made of the rate at which the plants are developing under these storage conditions and predictions are made as to the date on which the material will need to be subcultured.
3. Physiological status: Visual analysis is made of the physiological state of the material; for example, some clones begin to turn yellow after several months, whilst others begin to tuberize *in vitro* (Figure 5.2).

The results of all these observations are stored in a computer database. This computer database also contains other important information about these *in vitro* cultures. Figure 5.4 shows a typical data profile. Codes can be used to show date of introduction, status (i.e. propagation, conservation, risk of loss, lost, etc.), growth rate, predicted transfer date, location of duplicate cultures, etc. Although this type of database may not be necessary with a small collection, once there are several thousand accessions this information is a vital part of an integrated *in vitro* germplasm management system.

5.6 GENETIC STABILITY OF *IN VITRO* MATERIAL

The principal objective of a clonal germplasm collection is to maintain intact, specific gene combinations. If, after a period of storage, the genotype is not the same as the original material that was introduced, then the validity of the system must be questioned. Certain *in vitro* techniques such as callus and protoplast culture are known to induce varying degrees of genetic variability (Scowcroft and Larking, 1982; Thomas, 1981). Other methods such as meristem culture and micropropagation are said to be stable. The measure of stability is, however, only as good as the methods used to analyse it. Let us now look at some of the methods used to assess variability of *in vitro* material.

EXAMPLE

CENTRO INTERNACIONAL DE LA PAPA
In Vitro CULTIVATED COLLECTION

W LAB NUMBER	COLLECTION NUMBER	SPECIES	COUNTRY	STATUS	DATE L.T.S.	TRANSF INTERVL	DATE TRANSFER	GROW RATE	PHYSIOL CHARACT	DUPLICATE IN
3	CIP 701179	ADG	PER	5	11/09/84	24	11/86	5	FNSSY	E85G84
4	CIP 700852	STN	PER	5	02/19/85	24	02/87	9	MRRLG	E85
5	CIP 702286	ADG	PER	1		0		2	AARCY	–
6	CIP 700541	CHA	BOL	5	11/27/85	18	05/87	2	MRRLD	E85
7	CIP 701410	JUZ	PER	5	11/27/85	20	07/87	1	FNSLG	E85
8	CIP 700407	STN	PER	1		18		4	FTSSY	–
9	CIP 700605	ADG	BOL	5	05/11/86	18	11/87	8	MNRLG	E85
10	CIP 702328	ADG	ECU	5	02/19/85	20	10/86	3	AARXY	E85
11	CIP 701396	ADG	PER	5	11/09/84	24	11/86	9	MNSSG	E85G84
12	CIP 702775	ADG	PER	5	02/19/85	20	10/86	9	FRSLG	E85

CODES USED

STATUS: 1 = Introduced   2 = Lost   3 = Infected   4 = Propagated   5 = Long Term Storage

GROWTH RATE: Number × 10 in percentage (per year)

PHYSIOLOGIC CHARACTERISTICS: Five letters RTSPF for Roots, Tubers, Shoots, Plant Growth and Foliage

ROOTS

F = Few
M = Many
A = Many + aerial

TUBERIZATION

N = None
T = Tendency
R = Rapid
A = Rapid + abundant tubers

SHOOTS

S = Single
R = Ramified

PLANT GROWTH

S = Small plant, slow growth
L = Large plant, rapid growth
C = Small plant, senescent
Z = Large plant, senescent

FOLIAGE COLOR

Y = Yellowing
G = Green
D = Dark green

COUNTRY AND YEAR OF DUPLICATION: E = Ecuador   G = West Germany

Figure 5.4 A typical data profile sheet of *in vitro* accessions.

5.6.1 Morphological analysis

In experiments where genetic variability was specifically induced, variation was found in a wide range of morphological characters, e.g. leaf shape, flower size, flower colour, tuber skin colour. It is possible, therefore, to make a morphological analysis of plants after a period of *in vitro* storage. These techniques would not, however, show other genetic changes, i.e. changes in tuber proteins, and the morphological characters may be strongly affected by the environment. Another problem is that this type of analysis is very time-consuming and requires a lot of space and labour. It would not be a functional method for large-scale screening of an *in vitro* germplasm collection.

5.6.2 Protein and isoenzyme electrophoresis

In the identification of potato varieties, much emphasis is placed on the soluble protein electrophoresis of tuber storage protein (Stegemann, 1979; Desborough and Peloquin, 1969). The pattern of a genotype is distinct. We have observed that the protein electrophoresis patterns of physiologically mature *in vitro* induced tubers is similar to the pattern produced under *in vivo* conditions (Tovar and Dodds, unpublished). *In vitro* tubers can be rapidly induced and the material used for routine screening (Estrada *et al.*, 1986).

Protein electrophoresis does, however, also have some severe limitations. The electrophoretic pattern is a representation of certain gene products, but represents only a fraction of the total genome, and, in particular, proteins that are probably very well conserved. The protein pattern is also affected by the physiological state of the material, e.g. if physiologically immature tubers are analysed, the protein profiles are very different from those of mature tubers of the same genotype (Stegemann *et al.*, 1973).

Having indicated the shortcomings of this technique, we note that it is probably the most sensitive and reproducible technique available at the present time. Materials in the *in vitro* germplasm collection of the Potato Center are removed every two or three years, tubers are induced, either *in vitro* or *in vivo*, and protein profiles are prepared. The pattern is compared with the original tuber sample to see if any variation can be detected. To date in all samples analysed, no detectable differences have been identified in *in vitro* conserved material after over five years in storage. Of course, minor genetic changes may not have been detected. The Potato Center has initiated a research programme to study possible, genetic variation *in vitro* using restriction fragment length polymorphism (RFLP). This method will analyse the genes themselves and not the products of some of those genes.

5.7 DUPLICATION OF SAMPLES

With any germplasm collection there is always the question of mainte-nance of duplicate samples. It is hard if not impossible to put a finite value on a germplasm collection. It is somewhat easier to put a cost on an *in vitro* introduction (at least $6 US). It follows, therefore, that whenever feasible a duplicate set of *in vitro* cultures should be maintained. Duplicate copies of all CIP's *in vitro* materials are maintained outside Peru (in Ecuador and West Germany). This procedure further secures a valuable asset that, if lost, could never be replaced.

REFERENCES

Addicott, F. T. and Lyon, J. L. (1969) Physiology of abscisic acid and related substances. *Ann. Rev. Potato Physiol.*, **20**, 139–154.

Bajaj, Y. P. S. (1981) Regeneration of plants from potato meristem freeze preserved for 24 months. *Euphytica*, **30**, 141–145.

Bajaj, Y. P. S. and Reinert, J. (1977) in *Applied and Fundamental Aspects of Plant Cell Tissue and Organ Culture* (ed. J. Reinert and Y. P. S. Bajaj), Springer-Verlag, Berlin, pp. 757–776.

Beckman, J. S. and Soller, M. (1984) Restriction fragment length poly-morphisms in genetic improvement: Methodologies, mapping and costs. *Theor. Appl. Genet.*, **67**, 35–43.

Bernatzky, R. and Tanksley, S. D. (1986) Toward a saturated linkage map in tomato based on isozymes and random cDNA sequences. *Genetics*, **112**, 867–878.

Buckner, B. and Hyde, B. B. (1985) Chloroplast DNA variation between the common cultivated potato (*Solanum tuberosum* subsp. *tuberosum*) and several South American relatives. *Theor. Appl. Genet.*, **71**, 527–531.

Constabel, F. (1982) in *Plant Tissue Culture Methods* (ed. L. Wetter and F. Constabel), CNRC, Saskatoon, pp. 38–48.

Desborough, S. and Peloquin, S. J. (1969) Tuber proteins from haploids, selfs, and cultivars of group tuberosum separated by acid gel disc electrophoresis. *Theor. Appl. Genet.*, **39**, 43–47.

Dodds, J. H. (1986) in *Experiments in Plant Tissue Culture* (ed. J. H. Dodds and L. W. Roberts), Cambridge University Press, Cambridge, pp. 172–180.

Estrada, R., Tovar, P. and Dodds, J. H. (1986) Induction of *in vitro* tubers in a broad range of potato genotypes. *Plant Cell Tissue and Organ Culture*, **7**, 3–10.

Frankel, O. H. and Hawkes, J. G. (1975) *Crop Genetic Resources for Today and Tomorrow*, Cambridge University Press, Cambridge.

Glover, D. M. (1980) *Genetic Engineering Cloning DNA*, Chapman and Hall, pp. 48–66.

Grout, B. W. W. and Henshaw, G. G. (1978) Freeze preservation of potato shoot tip cultures. *Ann. Bot.*, **42**, 1227–1229.

Harmey, M. A., Crawley, M. P. and Clinch, P. E. M. (1966) The effect of growth regulators on the tuberisation of cultured stem pieces of *S. tuberosum. Eur. Potato J.*, **9**, 146–151.

Henshaw, G. G. (1979) Tissue cultures and germplasm storage. *IAPTC Newsletter*, **28**, 2–7.

Henshaw, G. G. and O'Hara, J. F. (1983) in *Plant Biotechnology* (ed. S. H. Mantell and H. Smith), Cambridge University Press, Cambridge, pp. 219–238.

Humphries, E. C. and Dyson, E. C. (1967) Effect of growth inhibitor *N*-Dimethylamino succinamic acid (B9) on potato plants in the field. *Eur. Potato J.*, **10**, 116–126.

IBPGR (1984) *The Potential for Using in Vitro Techniques for Collecting Germplasm*, IBPGR Secretariat, Rome, Italy.

Kanai, R. and Edwards, G. E. (1973) Purification of enzymically isolated protoplasts from C_3m C_4 and CAM plants using an aqueous dextranpolyethylene glycol two phase system. *Plant Physiol.*, **52**, 484–490.

Kartha, K. K. (1985) in *Cryopreservation of Plant Cells and Organs*, CRC Press, California, USA, pp. 116–134.

Kartha, K. K., Leung, N. L. and Gamborg, O. L. (1979) Freeze preservation of pea meristems in liquid nitrogen and subsequent plant regeneration. *Plant Sci. Lett.*, **15**, 7–15.

Larkin, P. J. and Scowcroft, W. R. (1981) Somaclonal variation – a novel source of genetic variability from cell cultures for improvement. *Theor. Appl. Genet.*, **60**, 197–214.

Maliga, P. (1978) in *Frontiers of Plant Tissue Culture 1978* (ed. T. A. Thorpe), IAPTC, pp. 381–392.

Morel, G. (1975) in *Crop Genetic Resources for Today and Tomorrow* (ed. O. H. Frankel and J. G. Hawkes), Cambridge University Press, Cambridge, pp. 327–332.

Murashige and Skoog (1962) Revised medium for rapid growth and bioassay with tobacco tissue cultures. *Physiol. Plant*, **15**, 473–497.

Nitzsche, W. (1983) in *Handbook of Plant Cell Culture* (ed. D. A. Evans, W. R. Sharp, P. V. Ammirato and Y. Yamada), Macmillan, New York, pp. 782–805.

Paupardin, C. and Tizio, R. (1970) Sur la présence de composès phenoliques dans les germes de pomme de terra cultives *in vitro. C. R. Acad. Sci. Ser. D Natur. (Paris)*, **269**, 1668–1670.

Roca, W. M., Espinoza, N. O., Roca, M. R. and Bryan, E. J. (1978) A tissue culture method for rapid propagation of potatoes. *Am. Potato J.*, **55**, 691–701.

Sadatz, W. and Standke, K. H. C. (1978) Untersuchungan Minimalwachstum von Kartoffeln *in vitro*. *Landbauforschung Volksnnode*, **28**, 75–78.

Scowcroft, W. R. and Larkin, P. J. (1982) in *Plant Improvement and Somatic Cell Genetics* (ed. J. K. Vasil, W. R. Scowcroft and K. J. Frey), Academic Press, pp. 159–178.

Silva, G. H. (1985) *In vitro* storage of potato tuber explants and subsequent plant regeneration. *HortScience*, **20**(1), 139–140.

Stecco, V. L. and Tizio, R. (1982) Cetion du CCC sur la tuberisation des germes de Pomme de terre cultivés *in vitro* caun milieu minéral dé pour vu de sucre. *C. R. Acad. Sci. Ser. III*, **294**, 901–904.

Stegemann, H. (1979) in *Conference on the Biology and Taxonomy of the Solanacea* (ed. J. G. Hawkes, R. N. Lester and A. D. Skelding), Linnean Society Symposium Series 7, pp. 279–284.

Stegemann, H., Francksen, H. and Macko, V. (1973) Potato proteins: genetic and physiological changes, evaluated by one- and two-dimensional PAA-gel-techniques. *Z. Naturforsach.*, **28c**(11–12), 722–732.

Thomas, E. (1981) Plant regeneration from shoot culture derived protoplasts to tetraploid potato. *Plant Sci. Lett.*, **23**, 81–88.

Westcott, R. J., Henshaw, G. G. and Grout, B. W. W. (1977a) Tissue culture methods and germplasm storage in potato. *Acta Hort.*, **68**, 45–49.

Westcott, R. J., Henshaw, G. G. and Roca, W. M. (1977b) Tissue culture storage of potato germplasm: Culture initiation and plant regeneration. *Plant Sci. Lett.*, **9**, 309–315.

Wilkins, C. P. and Dodds, J. H. (1983) The application of tissue culture techniques to plant genetic conservation. *Sci. Prog.*, **68**, 281–307.

Withers, L. A. (1980) *Tissue Culture Storage for Genetic Conservation*, IBPGR Technical Report AGP/80/8, Rome.

Withers, L. A. (1983) in *Plant Biotechnology* (ed. S. H. Mantell and H. Smith), Cambridge University Press, Cambridge, pp. 187–218.

Withers, L. A. and Davey, M. R. (1978) A fine structural study of the freeze preservation of plant tissue cultures. *Protoplasma*, **94**, 207–219.

Withers, L. A. and King, P. J. (1979) Proline – A novel cryoprotectant for the freeze preservation of cultured cells of Zea mays. *Plant Physiol.*, **64**, 675–678.

Withers, L. A. and Street, H. E. (1977) in *Plant Tissue Culture and its Biotechnological Applications* (ed. W. Barz, E. Reinert and M. H. Zenk), Springer-Verlag, Berlin, pp. 226–244.

Yidana, J. A., Withers, L. A. and Ivins, J. D. (1990) Development of a simple method for collecting and propagating cocoa germplasm *in vitro*. *Acta Hort*. (in press).

6

Conservation *in vitro* of cassava germplasm

L. VILLEGAS and M. BRAVATO

6.1 INTRODUCTION

Cassava (*Manihot esculenta* Crantz) is one of the cheapest sources of calories for human consumption in the tropical lowlands, with the highest potential production of calories per hectare and per year (De Vries *et al.*, 1967). It has grown in northern Amazonia for many years, and from there it was introduced into the Caribbean, Africa and Asia. It grows in areas with rainfall seasons of 6–8 months per year and with a minimum of 750 mm rain/year. The crop can survive up to six-month drought periods by reducing new leaf formation and becoming essentially dormant until the onset of the rain, when root reserves are used to produce new foliage (Cours, 1951). In the tropics, between latitudes of 30°N and 30°S and at altitudes below 2000 m above sea level, the temperature is usually higher than 20°C, and cassava grows well in these conditions. Day length is also adequate. Shading for periods of up to two months has only limited effects on plant development and root growth. Cassava has no special soil requirements, growing in low-pH and poor-fertility soils, in which other crops cannot be cultivated. In these old and most depleted soils, cassava is able to produce reasonable root yields. Only high salinity, alkalinity and water excess represent severe restrictions for cassava production. Low nutrient levels reduce root growth less in cassava than in other crops, the yield obtained being due to the long root-filling period rather than to a maximal crop growth rate of the plant. Additionally, cassava plants can be left in the field for long periods, being collected at the moment of root usage. For these reasons, in the last 20–30 years, some efforts have been dedicated to preserve the available genetic variability of cassava germplasm.

Cassava is a crop vegetatively propagated through the use of stem

In Vitro *Methods for Conservation of Plant Genetic Resources. Edited by John H. Dodds.*
Published in 1990 by Chapman and Hall, London.   ISBN 0 412 33870 X

cuttings (stakes). Plants have to be replanted in the fields every year with the risk of accumulating and propagating new diseases. This makes necessary the maintenance of plant germplasm under conditions in which individual genotypes can be preserved unchanged at minimal cost. Therefore, certain tissue-culture techniques have been used for germplasm conservation.

The technique for regenerating cassava plants from shoot apical meristems was developed by Kartha *et al.* (1974) and extended to the preservation of germplasm by Roca (CIAT, 1980) at the International Center of Tropical Agriculture (CIAT) in Colombia, where the most extensive collection *in vitro* of cassava germplasm has been assembled. Other valuable efforts in cassava germplasm conservation have been made in the Centro Nacional de Recursos Genéticos (CENARGEN), Brazil; the International Institute of Tropical Agriculture (IITA) in Nigeria; the National Research Council of Canada; the Punjab Agricultural University, India; and the national programmes of China, India, Indonesia, Malaysia, and Thailand. Another collection exists in our laboratory at the Instituto Internacional de Estudios Avanzados (IDEA), Venezuela.

Micropropagation in standard conditions is the most widely used tissue-culture technique for cassava germplasm conservation. However, in these conditions the frequent transfer makes the technique costly and increases the risk of contamination. Besides, positive selection during subculturing could be a source of genotypic variation. Consequently, it is necessary to reduce the rate of growth so that limited attention is required for plantlets in storage. Cryopreservation of plant material and establishment of cultured cell lines also seem to be feasible ways of extending tissue-culture techniques to long-term storage of genetically valuable material.

6.2 CASSAVA GERMPLASM COLLECTION

All *in vitro* methods require the collection of material from the field. The conventional method for material collection was basically described by the International Board for Plant Genetic Resources (1983).

For cultivated or wild cassava, germplasm collection in the form of stakes, the normal propagation material, is the most appropriate. Plants to be sampled in the field should be nearly mature to mature if possible. Plants are pulled out before cutting any branches to inspect the roots for symptoms of 'frog skin' disease, which should not be introduced into the germplasm maintenance area. Also, the cutting of stakes from plants affected by bacteriosis, superelongation and virus diseases must be avoided. A few stakes about 50 cm long are obtained from each plant. The stakes must be treated with a fungicide–insecticide mixture. When

112

meristem cultures are to be used, the stakes collected should be covered at the upper end with parafilm to prevent desiccation. The stakes are planted in small pots and the sprouts obtained are used as a source of meristems. A true identification of the material is essential. Material from the same clone should be labelled, indicating the location, date of collection and common name. A collection code could be assigned.

Alternative *in vitro* methods for collecting material in the field, based on *in vitro* standard laboratory techniques, have been developed. Withers (1987) reports some of these methods used in cassava. One of them involves collecting buds in the field and taking them to a nearby building no more than 3–4 hours' drive away. By working in a clean area and using an alcohol burner to flame the instruments and the culture vessels, the buds are disinfected and surface-sterilized. The shoot apices are dissected, using a magnifying lens, and then inoculated into a culture medium in which a piece of filter paper containing dry antibiotic is inserted. A contamination rate of 5–8% has been found when using this procedure to transfer material from Sumatra to CIAT (Withers, 1987).

Another method is carried out completely in the field, where shoots are excised, disinfected, surface-sterilized, dissected and inoculated into medium. An antibiotic and a fungicide are added via a piece of filter paper. In this case, a clean table swabbed with diluted bleach and a three-sided box that provides some protection from dust are used. The utilized instruments are disinfected with bleach or with an alcohol burner.

Starting from the collected stakes, standard laboratory *in vitro* culture initiation techniques are used. Initiating an *in vitro* culture involves: selecting and trimming to size a suitable explant; washing and surface disinfection; surface sterilization; washing with sterile water; and dissection of the explant to be inoculated into the culture medium. The most commonly used explant is the shoot apical meristem.

6.3 MERISTEM CULTURE

Cassava meristem culture was developed successfully by Kartha *et al.* (1974). Combined with thermotherapy, it makes it possible to establish and maintain *in vitro* cassava collections free of contaminating micro-organisms (Roca *et al.*, 1982; Schilde-Rentscheler and Roca, 1983, 1987; Roca, 1985; Villegas and Bravato, 1988).

Preliminary results on the phenotypic and genotopic stability of plants propagated by this technique have been obtained. The evaluation of qualitative and quantitative characteristics and electrophoretic analysis have so far shown that the plants maintain their clonal characteristics (Centro Internacional de Agricultura Tropical, 1989).

The method developed for meristem culture in cassava from stakes collected in the field is as follows. The stakes, previously treated with a fungicide–insecticide mixture, are planted in a disinfected soil mixture and placed in a thermotherapy chamber at 40°C during the day and 35°C during the night. Twelve-hour photoperiods with 3000 lux light intensity were used. Two weeks later, buds of sprouted cuttings, as well as subsequent secondary buds developed during a period of 30–45 days, are used to isolate the apical meristem. The outer whorls of visible leaves should be removed before surface-sterilization steps. By working in a sterile environment provided by a laminar air-flow cabinet, the buds are surface sterilized in 70% ethanol and 0.125% sodium hypochlorite for 3 min and then rinsed thoroughly three times with sterile distilled water. Meristems measuring 0.4 mm in length with one or two leaf primordia are isolated under a stereo microscope, transferred to a culture medium and incubated in a controlled environment consisting of 26–28°C temperature, 12 h photoperiod and 2000 lux light intensity to promote organ differentiation and plant growth (Roca, CIAT 1980; Villegas and Bravato, 1988).

The culture medium used by Kartha *et al.* (1974) consists of macro- and microelements according to Murashige and Skoog (1962), vitamins as in B5 medium (Gamborg *et al.*, 1968) and 2% sucrose. Growth regulators – benzyladenine (BA), naphthalene acetic acid (NAA) and giberellic acid (GA3) – are incorporated at 0.1 ppm, 0.2 ppm and 0.03 ppm respectively. Later, Roca reported the existence of a pronounced effect of the cultivar on the meristem culture and proposed some modifications in the growth regulators and vitamins. Proposed concentrations are: 0.05 ppm BA; 0.02 ppm NAA; 0.05 ppm GA3; 0.4 ppm thiamine; and 100 ppm m-inositol (Centro Internacional de Agricultura Tropical, 1980). Recently new changes were introduced, BA going from 0.05 ppm to 0.04 ppm and thiamine from 0.4 ppm to 1 ppm (Roca, personal communication).

In good growth conditions, the meristem increases in size and becomes greenish in colour. The shoot differentiation starts within 7–10 days, and is followed by root formation. At this stage, it is advisable to transfer the explant to a fresh medium. The development of a complete plant, 5–7 cm in height, occurs within 5–6 weeks.

Another method recently developed to obtain disease-free plants involves the use of *in vitro* thermotherapy (Centro Internacional de Agricultura Tropical, 1987). In this case the vegetal material to be used consists of *in vitro* plants obtained either from buds of sprouted cuttings of stakes collected in the field which have not been treated with heat, or from buds collected *in vitro* in the field as described in cassava germplasm collection. *In vitro* thermotherapy consists of the following steps: (1) propagation of the *in vitro* plant using shoot apices 0.5 cm in length and single node cuttings with their respective axillary bud grown for 12 days under thermotherapy treatment (40°C day, 35°C night; 12 h photoperiods, 2000–3000 lux); (2) cutting shoot apices of the same size from the propagated plants and again applying thermotherapy for 12 days. Cutting and thermotherapy are repeated one more time.

The culture medium used for the explants in each step is the one described for meristem culture. The regenerated plants are propagated for virus testing. As previously stated, using the plants regenerated *in vitro* and indexing them for known diseases makes it possible to establish collections of pathogen-free *in vitro* cassava germplasm.

Two different collection types could be established: one devoted to the production of plants to be tested and used for production in the field, active collection, and another for the maintenance of the germplasm resource.

The first type of collection requires the production of an appreciable number of plants of one or more cultivars to be transferred to the field after adequate multiplication and adaptation under controlled conditions. In this case, the initial material, mother plants, originate from plants collected in the field and cleaned, and meristem-cultured and regenerated in the laboratory, as well as from plants received from the collections to be described in the maintenance collection. A normal growth rate of the plants is desired and frequent replanting of the material must be done. An example of this type is the laboratory of IDEA (Villegas and Bravato, 1988).

The second type, maintenance collection, is used to store a large number of well-differentiated cultivars in order to keep as constant as possible the genetic base. For this type of collection, with relatively few plants of each cultivar, it is necessary to reduce the growth rate of the plants to avoid the frequent replanting of the material with the potential risks of contamination and genetic variation. An excellent example of this type is the Elite Cassava Germplasm collection maintained at CIAT (Centro Internacional de Agricultura Tropical, 1983,1984; Schilde-Rentschler and Roca, 1987).

6.4 NORMAL GROWTH-RATE CONSERVATION

This technique is based on the storage of shoot-tip culture or meristem-derived plants under normal growth conditions. It requires that the plants be transferred to a fresh medium before culture deterioration takes place.

Subculture of the plants is carried out as follows. Plants are removed from the test tubes and cut into single nodes. Each nodal cutting is transferred to a medium like the one used for meristem culture. In some cultivars tested in our laboratory, removal of BA from the medium increases development of the plants (Zypman, unpublished). The apical bud also obtained is cultured in a second medium proposed by Roca (CIAT, 1980) to improve root formation. This medium contains: one-third of the macro- and microelements from MS; 0.01 ppm NAA; 1 ppm thiamine; 100 ppm m-inositol; and 25 ppm of 10-52-10 fertilizer. After six weeks of incubation, the plants will be ready for either subculture again or for transfer to soil in controlled conditions, before being used for field assays. In each subculture process, the material is multiplied by a factor of 5. This process can be repeated as many times as required to produce the desired number of plants. If the plants are not used for field assays, the incubation period can be extended, thus increasing the propagation rate. Our experience in the laboratory indicates that most of the cultivars maintained under normal growth conditions can last up to six months before being transferred to a fresh medium.

The normal growth conditions used were 26–28°C and light intensity 2000–3500 lux, increasing during a period of 0–6 weeks.

Our laboratory is located near Caracas city, approximately 10°30′N, and 1400 above sea level. The mean year temperature is 18°C, with a minimum of 10°C in January and a maximum of 26°C in August–September. To obtain the desired normal growth conditions a conservation room was built, consisting of a concrete structure 5 m in length, 2.5 m in width, and 4 m in height. This room has a capacity of 20 000 plants in 18 × 150 mm test tubes. Twelve units of five panel shelves, painted white, were used to store the tube racks. One, two or three daylight-type 40 W fluorescent tubes were fixed at the bottom of the panels. The distance between the tubes and the next panel was 30 cm. The light intensity at the surface of the panels was 2500 lux, 3500 lux and 4500 lux. The temperature was established by the dissipation of heat from the light tubes during the day and from six thermal electrical resistances of 600 W each, at night, in conjunction with two air-conditioning units of 6000 BTU each. The photoperiod was controlled by an electric clock and a relay with sufficient capacity to change from the lights to the electrical resistances. No other special installation was required.

6.5 REDUCED GROWTH-RATE CONSERVATION

For cassava, long-term storage has been made possible basically by reducing temperature and illumination, and also introducing minor modifications into the culture media. The following conditions are used at CIAT to maintain the *in vitro* collection: temperature of 23°C ± 1°C; photoperiod of 12 h with 600–800 lux of light intensity. The explants to be used are apical buds of *in vitro* plants cultured in a medium containing the macro- and microelements from MS; 2% sucrose; 0.02 BA; 0.1 ppm GA3; 0.01 ppm NAA; 1 ppm thiamine; 100 ppm m-inositol. The test tubes are then covered with aluminium foil. These conditions increase the transfer interval up to 12–18 months (Central Internacional de Agricultura Tropical, 1987).

6.6 CRYOPRESERVATION

The basic principles are covered elsewhere in this book. Here we will deal with the specific case of cassava.

The method described by Bajaj is basically as follows. The meristems were precultured over 5–7 days in MS; 0.1 ppm BA; 0.2 ppm NAA; 0.05 ppm GA3; and dimethylsulphoxide. The selected cultures showing signs of growth were placed in filter paper soaked in a mixture at 5% each of dimethylsulphoxide, sucrose and glycerol and kept for 2–4 h. The meristems were frozen along with the filter paper wrapped in aluminium foil. The freezing was done by direct immersion in liquid nitrogen or by first exposing the samples to liquid nitrogen vapour and then gradually immersing the samples in liquid nitrogen, in order to induce a slow freezing. The material was stored in liquid nitrogen for different periods, after which the meristems were thawed in warm water at 35–40°C and cultured in MS, BA, NAA, and GA3 medium. The survival of the retrieved material was judged according to the following criteria: increase in size; turning to green; proliferation to form calluses and to develop into shoots and plantlets. Using this method, Bajaj (1985) claims to have obtained up to 29.3% of survival as compared to the control unfrozen tissues.

The method described by Kartha et al. (1982) is essentially as follows. Meristems were dropped into hormone-free MS liquid medium, and this was gradually diluted with an equal volume of 30% v/v dimethylsulphoxide in the MS medium containing 3% sucrose. The meristems were kept in this freezing solution for 30 min. A piece of aluminium foil was placed inside a plastic Petri dish and the meristems were distributed over the foil along with 2–3 µl of the freezing solution. The covered Petri dishes were transferred to the freezing chamber of a Cryo-Med 1000 Biological Freezing System. The cooling rate was monitored in one of the droplets. A rate of 0.5°C/min was used to reach the terminal freezing temperatures of −20°C, −25°C, −30°C and −40°C. Other dishes containing meristems were completely immersed in liquid nitrogen for 1 h. Later, the aluminium foils with the meristems were removed from the Petri dishes and thawed for 10 min by immersion in hormone-free MS medium at 37°C. The meristems were then removed and cultured as previously described by Kartha et al. (1974) and Kartha and Gamborg (1975). With this droplet-freezing method at a cooling rate of −0.5°C/min, the viability of the meristems depends upon the terminal freezing temperature. At −25°C over 90% of the meristems survived and over 60% of them differentiated into plantlets. At −20°C it is possible that there was a failure in the total phase change, and at −30°C and −40°C the meristems showed a further loss in viability.

When meristems frozen at −25°C were immersed in liquid nitrogen, a great variability was observed in their morphogenic response. Kartha *et al.* (1982) found that 11 plantlets were obtained from 186 meristems initially frozen. From meristems frozen to −30°C and −40°C the ratios reported are 4/264 and 2/139 plantlets per meristem frozen, respectively. It is apparent that the initial difference due to final-freeze temperature disappeared when meristems were immersed in liquid nitrogen. Better relations were obtained using the survival criteria proposed by Bajaj (1983). This suggests the possibility of obtaining growth in callus formation, without plantlet regeneration, due to partial survival of meristem tissues.

From the experience of the two laboratories working in the same area of cryopreservation, the facts to be considered for the purpose of improving the procedures are: preculture; use of adequate cryopreservation mixture; rate of temperature reduction; homogeneity in cooling the sample; final storage temperature; and rate of thawing.

6.7 DEVELOPMENT OF NEW METHODS

Plant cell cultures have been considered to be ideal tools for a variety of biotechnological applications and novel ways of utilizing plant genetic resources. Therefore, the importance of storing cell cultures has been recognized. Freeze-preservation is at present the technique available for cell culture, being the most appropriate option, since no genetic changes occur with time (Withers, 1984). In addition, this culture system has been considered more amenable to cryopreservation than meristems or shoot tips (Kartha, 1985). However, the number of species preserved by this method is low, when compared with the vast range of species currently under general investigation. Consequently, research must continue towards the improvement of the methodology for storing cell cultures.

The previous establishment of homogeneous cell lines is required. Some work on cassava protoplasts, as a way to regenerate plants, has been reported (Shahin and Shepard, 1980; Szabados *et al.*, 1987). However, no reports have been published on cell lines. Experiments to create cassava cell lines were tried, with isoelectric focusing of protoplasts, in our laboratory (Villegas and Bravato, 1987; Santana, 1989), using the technique described by Griffing *et al.* (1985). However, only cell colonies were obtained and no plant regenerated or cell was evaluated. Consequently, no homogeneity of the population could be claimed. We certainly believe that further research in this area would be a help towards the use of a variety of techniques of germplasm utilization.

REFERENCES

Bajaj, Y. P. S. (1977) Clonal propagation and cryopreservation of cassava through tissue culture. *Crop. Improv.*, **4**, 198–204.

Bajaj, Y. P. S. (1983) Cassava plants from meristem cultures freeze-preserved for three years. *Field Crops Res.*, **7**, 161–167.

Bajaj, Y. P. S. (1985) Cryopreservation of germplasm of potato (Solanum tuberosum L.) and cassava (Manihot esculenta Crantz): Viability of excised meristems cryopreserved up to four years. *Ind. J. Exp. Biol.*, **26**, 285–287.

Centro Internacional de Agricultura Tropical (1980) *El Cultivo de Meristemas de Yuca*, CIAT, Cali, Colombia.

Centro Internacional de Agricultura Tropical (1983) *Elite Cassava Germplasm from CIAT*, CIAT, Cali, Colombia.

Centro Internacional de Agricultura Tropical (1984) *El Cultivo de Meristemas para la Conservación de Germoplasma de Yuca 'in vitro'*, CIAT, Cali, Colombia.

Centro Internacional de Agricultura Tropical (1987) *Biotechnology Research Unit. Annual Report 1986*, CIAT, Cali, Colombia.

Centro Internacional de Agricultura Tropical (1989) *Biotechnology Research Unit. Annual Report 1988*, CIAT, Cali, Colombia.

Cock, J. H. (1985) *Cassava, New Potential for a Neglected Crop*, Westview Press, Boulder, USA.

Cours, G. (1951) Le manoic a Madagascar. *Mem. Inst. Sci. Madagascar*, **38**, 203–400.

De Vries, C. A., Ferwerda, J. D. and Flach, M. (1967) Choice of food crops in relation to actual and potential production in the Tropics. *Neth. J. Agri. Sci.*, **15**, 241–248.

Gamborg, O. L., Miller, R. A. and Ojima, K. (1968). Nutrient requirement of suspension cultures of soybean root cells. *Exp. Cell Res.*, **50**, 151–158.

Griffing, L. R., Cuttler, A. J., Shargool, P. D. and Fowke, L. C. (1985). Isoelectric focusing of plant cell protoplasts. *Plant Physiol.*, **77**, 765–769.

Henshaw, G. G., O'Hara, J. F. and Westcott, R. J. (1980). In *Tissue Culture Methods for Plant Pathologists* (ed. D. S. Ingram and J. P. Helgeson), Blackwell Scientific Publications, Oxford, pp. 71–76.

Hershey, C. H. (1987) in *Cassava Breeding: A Multidisciplinary Review* (ed C. H. Hershey), CIAT, Cali, Colombia, pp. 1–24.

International Board for Plant Genetic Resources (1983) *Genetic Resources of Cassava and Wild Relatives*, IBPGR Secretariat, Rome, Italy.

International Board for Plant Genetic Resources (1986) *IBPGR Advisory Committee on 'in vitro' Storage. Report on the Third Meeting*, IBPGR, Rome, Italy.

Kartha, K. K. (1985) Cryopreservation of plant cells and organs. *Newsletter IAPTC*, **45**, 2–15.

Kartha, K. K. and Gamborg, O. L. (1975) Elimination of cassava mosaic disease by meristem culture. *Phytopathology*, **65**, 826–828.

Kartha, K. K., Camborg, O. L., Constabel, F. and Shylux, J. P. (1974) Regeneration of cassava plants from shoot apical meristems. *Plant Sci. Lett.*, **21**, 107–113.

Kartha, K. K., Leung, N. L. and Mroginski, L. A. (1982) *In vitro* growth responses and plant regeneration from cryopreserved meristems of cassava (Manihot esculenta Crantz). *Z. Pflazenphysiol. Bd.*, **107–S**, 133–140.

Murashige, T. and Skoog, F. (1962) A revised medium for rapid growth and bioassays with tobacco tissue culture. *Physiol. Plant.*, **15**, 473–497.

Roca, W. M. (1985) in *Biotechnology in International Agricultural Research: Proc. Inter-center Seminar on IARCs and Biotechnology 1984*, IRRI, Manila, Philippines, pp. 3–10.

Roca, W. M., Rodriguez, J., Beltran, J., Roa, J. and Mafla, G. (1982) in *Plant Tissue Culture 1982* (ed. A. Fujiwara), International Association of Plant Tissue Culture, Tokyo, Japan, pp. 771–772.

Santana, M. A. (1989) Obtención y Caracterización de una Población de Protoplastos de Yuca (Manihot esculenta Crantz) Homogénea en cuanto a la Carga de Superficie de sus Membranas. Thesis, Universidad Simón Bolivar, Caracas, Venezuela.

Schilde-Rentscheler, L. and Roca, W. M. (1983) in *Global Workshop on Root and Tuber Crops Propagation* (ed J. H. Cock), CIAT, Cali, Colombia, pp. 89–93.

Schilde-Rentscheler, L. and Roca, W. M. (1987) in *Biotechnology in Agriculture and Forestry* (ed Y. P. S. Bajaj), Vol. 3, Springer-Verlag, Berlin, Heidelberg, pp. 453–465.

Shahin, E. A. and Shepard, J. F. (1980) Cassava mesophyll protoplasts: Isolation, proliferation and shoot formation. *Plant Sci. Lett.*, **17**, 459–465.

Szabados, L., Narvaez, J. and Roca, W. M. (1987) *Techniques for Isolation and Culture of Cassava (Manihot esculenta Crantz) Protoplasts*, Working Document No. 23, Biotechnology Research Unit, CIAT, Cali, Colombia.

Villegas, L. and Bravato, M. (1987). Separation of leaf protoplasts by isoelectric focusing. *In vitro*, **23**, 49A.

Villegas, L. and Bravato, M. (1988) Cultivo de células y tejidos para el mejoramiento de la yuca en Venezuela. *Interciencia*, **13**, 121–127.

Villegas, L., Bravato, M. and Zapata, C. (1988) in *Cultivo de Tejidos Vegetales Aplicado a la Producción Agricola* (ed. L. Villegas), Corporacion Andina de Fomento (CAF), Caracas, Venezuela, pp. 41–55.

Wilkins, C. P. and Dodds, J. H. (1983) The application of tissue culture techniques to plant genetic conservation. *Sci. Prog. Oxf.*, **68**, 259–284.

Withers, L. A. (1980) *Tissue Culture Storage for Genetic Conservation*, IBPGR Tech. Report, FAO/United Nations, Rome.

Withers, L. A. (1984) in *Cell Culture and Somatic Cell Genetics of Plants*, Vol. 1 (ed. I. K. Vasil), Academic Press, Orlando, Fl., pp. 608–620.

Withers, L. A. (1987) *In vitro* methods for collecting germplasm in the field. FAO/IBPGR. *Plant Genetic Resources Newsletter*, **69**, 2–6.

7

Conservation and distribution of sweet potato germplasm

C. G. KUO

7.1 INTRODUCTION

Sweet potato (*Ipomoea batatas* Lam.) is a member of the morning glory or bindweed family, the Convolvulaceae. It is a perennial herbaceous plant, but is generally grown and harvested in 3–5 months. An extensive, fibrous, adventitious root system is produced from the nodes of vines. Enlarged roots serving as a storage organ and variable in shape, size, number, skin colour (white, yellow, brown, purple), and flesh colour (white, yellow, orange, purple) are derived from secondary thickening of some adventitious roots. Storage roots are used mainly for human consumption in most countries. Sweet potato has tremendous yield potential and high nutrient productivity, and can survive a wide range of adverse environments. The crop can be a rich source of energy as well as of vitamins A and C. Small portions of production are used as feed or processed for starch and alcohol (Bouwkamp, 1984). Sweet potato for human consumption comes in the form of dessert, snacks or supplementary food, but is a staple food in Papua New Guinea and some other Oceanic countries. Young shoots, on the other hand, are also consumed in some tropical countries as a green vegetable; the terminal tips, petioles and tender leaves are eaten. Sweet potato greens are rich in vitamins, minerals and proteins, and are considered as an important nutritious food source in the tropics and subtropics, especially under adverse conditions (Villareal *et al.*, 1982).

It is generally accepted that sweet potato originated from Central America or northern South America, based on distribution patterns of wild relatives, variation in cultivated populations from America, and archaeological relics. Sweet potato is an important crop, ranking seventh in terms of worldwide production – 111 metric tons from eight million hectares in

In Vitro *Methods for Conservation of Plant Genetic Resources. Edited by John H. Dodds.* Published in 1991 by Chapman and Hall, London.   ISBN 0 412 33870 X

1985 (Horton, 1988). Asia accounts for 91% of the world's total production; China is the main producer, accounting for about 81% of the world's total production. There has been a significant reduction in the world's production area in recent years.

Sweet potato is grown from 48°N to 40°S latitudes (Anonymous, 1982; Bouwkamp, 1984). On the equator it is grown from sea level to 3000 m. Its growth is maximal at temperatures above 25°C; when temperatures fall below 12°C or exceed 35°C, growth is retarded. Dry-matter production increases with increasing soil temperatures from 20°C to 30°C, but declines beyond 30°C. Sweet potato is a sun-loving crop, and grows best with a well-distributed annual rainfall of 600–1600 mm. It is relatively drought-tolerant, mainly because of its potential for regeneration and root penetration.

Sweet potato plants rarely flower under tropical conditions. Even if they do produce seeds which can be stored for long periods of time, these are frequently heterogeneous and their regenerated plants do not reproduce original clones. This is one of the reasons why vegetative propagation is needed. Propagation by seeds is reserved only for breeding purpose to produce new clones. Vine cuttings about 30 cm from the tip are generally used for propagation, but sometimes cuttings from the middle vine portion are also used. In the areas where the plant cannot grow year-round, sprouts from storage roots of the previous crop which have been stored are used as planting materials. These methods are subject to disease infection, and are thus unsuitable for producing high yields with good quality and preserving disease-free planting stocks.

Sweet potato is attacked by a number of pests. In the humid tropics, scab (*Elsinoe batatas*) is the most prevalent disease, followed by *Fusarium* wilt (*Fusarium oxysporum*) and witches' broom caused by mycoplasm-like organisms. Soil rot (*Streptomyces ipomoea*), black rot (*Ceratocystis fimbriata*), Java black rot (*Diplodia tubericola*), scurf (*Monilochaetes infuscans*) and various virus diseases occur in sweet potato but their distribution and importance vary with region. In addition, sweet potato weevil (*Cylas formicarius*) is the most destructive insect in tropical Asia and Oceania.

The narrow genetic base of commonly grown sweet potato clones attests the need to introduce new sources of genetic diversity for use in the sweet potato improvement programme. According to the International Board for Plant Genetic Resources (IBPGR), there were 69 000 accessions of germplasm maintained worldwide; however, many of them are duplicated and it is estimated that about 3500 are true clones. Intensive collection activities sponsored by the IBPGR are in progress in some areas (De la Puente, 1988). Extensive collections of germplasm are maintained at the Asian Vegetable Research and Development Center (AVRDC) in Taiwan, the Centro de Agricultura de Investigacion y Ensenanza in Costa Rica, the International

Institute of Tropical Agriculture in Nigeria, and the International Potato Center (CIP) in Peru. In Asia and Oceania, many accessions are also maintained by individual national programmes; Papua New Guinea has the largest collection, with more than 2500 accessions (Takagi, 1988).

In order to maintain the genotypic integrity of these sweet potato germplasms, they must be propagated clonally and maintained in vegetative form at the repositories. Clonal propagation by cuttings and/or storage roots, and storage of sweet potato by traditional means, are time- and space-consuming and involve substantial costs. Under present circumstances, because of the vegetative nature of propagation, the sweet potato germplasm must be replanted every year. From the point of view of germplasm conservation, these field gene banks are, in most cases, unrepresentative of the range of genetic variability within the gene pool and most of them do not constitute more than a fraction of the variability which should be conserved for the future. Problems also exist in moving vegetative materials; survival of vine cuttings is low, and storage roots are bulky. Therefore, the IBPGR recommended the organization of a worldwide system of germplasm collection with *in vitro* gene banks (IBPGR, 1988).

In order to make an international germplasm system for sweet potato feasible, only pathogen-free material must be transported internationally. To meet this demand, an alternative method using tissue culture has become the principal technique. This method could be very effective in maintaining large populations of genetically pure, and disease-free, materials in a relatively small space, with reduced costs, until needed. Rapid multiplication of genetically pure material that is disease-free will also be of great value to production; it could lead to cheaper produce for the consumer. Since sweet potato is hexaploid and often difficult to induce to flower, once a system for plant regeneration from tissue and/or cell culture has been developed, that system might be used to help in clonal improvement.

The study of tissue-culture methods on sweet potato has not been intensive, despite its great potential for germplasm conservation and clonal improvement. The last review on this subject matter was by Henderson *et al.* (1983). The present chapter highlights the main developments in the subject matter since then. The emphasis, rather than being on tissue-culture methods *per se*, is placed on the regeneration of plantlets from sweet potato explants and the application of tissue culture to conservation and distribution of sweet potato germplasm.

7.2 *IN VITRO* CULTURE

The application of tissue-culture technology to the conservation and distribution of sweet potato germplasm depends upon the ability to successfully regenerate plants from small organs or pieces of tissues, which are called explants, of mature plants.

A number of primary explants have been used in the regeneration of small plants or 'plantlets' under *in vitro* conditions with varying degrees of success (Table 7.1). Sterilized explants are usually transferred to and cultured in solidified medium with 0.7–0.8% agar. Several basal nutrient media (Miller, MS, White) have been used, but the most widely used is the MS medium or its modification, which may be used at full strength or at half dilution. Light intensities of the order of 1000–5000 lux on a light–dark schedule of 16/8 hours are widely used. Most evidence indicates that the optimum temperature is in the range 22–28°C.

Usually, the first process undergone by explanted tissues is the formation of unorganized tissue (callus) at the cut surface; this is a natural response to wounding in many plants. Sweet potato plantlets can be regenerated: (1) through direct shoot development from pre-existing meristems (primordia) of explanted apical and/or lateral buds with little or no callus formation; and (2) through organogenesis or somatic embryogenesis of callus derived from explanted storage root tissue, leaf tissue, petiole tissue, stem tissue, cotyledon tissue or anthers or directly upon these explanted tissues (Table 7.1). The origin of the explanted tissue can be critical in the callus formation and/or the regeneration potential (Templeton-Somers and Collins, 1986a), which has been attributed to organized structures and/or anomalous cambia within the explant (Carswell and Locy, 1984; Hwang *et al.*, 1983). As far as the anther culture is concerned, the callus is originated from the anther wall, so all plantlets regenerated are diploid (Sehgal, 1978).

The size, growth duration and texture of callus varies according to treatments and clones. When given appropriate plant hormone(s) it is possible to regenerate plantlets from calli through organogenesis; this is *de novo* formation of meristematic loci (primordia), leading to the organization of well-defined shoot and/or root meristems which eventually develop into adventitious shoots and/or roots (Hwang *et al.*, 1983). The primordia may be formed from cells on the surface, or in the interior, of the callus tissues. The determination of shoot/root formation is often dependent on the ratio of plant hormone(s) in the nutrient medium. In most cases, roots are most prolific on the medium containing auxins such as 2,4-D(2,4-dichlorophenoxy acetic acid), IAA (indole-3-acetic acid), IBA (indole-3-butyric acid) and NAA (naphathalene acetic acid) (Chen, 1981; Carswell and Locy, 1984), and shoot formation is greatest on the medium containing cytokinins such as BA (benzylaminopurine), kinetin and zeatin (Gunckel *et al.*, 1972; Litz and Conover, 1978; Murata and Miyaji, 1986). At times, GA$_3$ (Giberellic acid) or ABA (abscisic acid) also bring about differentiation and organ formation on the callus (Gunckel *et al.*, 1972;

Table 7.1 Summary of *in vitro* plantlet regeneration of sweet potato from various explants

Explant	Clone	Callus	Organogenesis	Regeneration	Reference
			Medium for		
Cotyledon	Ningyuan Sanshirizao × Huabei 52–45 Ning 404-16 × Nonglin 26	MS 2,4-D 0.1 mg/l	MS	1/2 MS AD 15 mg/l Kin 0.5 mg/l	Xue (1987)
Leaf pieces	Tainung 57 Tainan 15 Red Tip Okinawa 100	MS 2,4-D 0.5 mg/l (or IBA 1 mg/l) or NAA 0.5 mg/l Kin 0.1 mg/l	MS IAA 1–10 mg/l (or IBA 1 mg/l) (or NAA 0.5 mg/l)	MS	Chen (1981)
	Jewel Caromex	MS BA 10 mg/l NAA 1 mg/l	MS BA 0.1 mg/l NAA 1 mg/l		Carswell and Locy (1984)
	White Star GaTG3	MS 2,4-D 0.5–4.0 mg/l	MS 2,4-D 2.0 mg/l CW 20% Kin 2 mg/l		Liu and Cantliffe (1984)
	Jewel		MS IAA 2 mg/l		Templeton-Somers and Collins (1986b)
	331 12-17 Qinnanhong	MS BA 0.01–0.1 mg/l NAA 2 mg/l	1/2 MS IAA 0.2–1.0 mg/l Kin 0.5–2.0 mg/l		Xin and Zhang (1987)
	Duclos XI	MS IAA or NAA 1 mg/l	KM Zeatin 2 mg/l	KM Zeatin 0.25 mg/l	Sihachakr and Ducreux (1987a)

Table 7.1 Continued

Explant	Clone	Callus	Organogenesis	Regeneration	Reference
			Medium for		
	K 221 (2X) DJ-13-2	MS 2,4-D 0.1–0.5 mg/l BA 0.1–0.5 mg/l	MS 2,4-D 0.1–0.5 mg/l BA 0–3 mg/l		Suga and Irikura (1988)
Petiole	Centennial Redmar	MS BA 2 mg/l NAA 0.1 mg/l	MS BA 2 mg/l NA 0.1 mg/l		Hwang et al. (1983)
	331 12-17 Qinnanhong	MS BA 0.01–1.0 mg/l NAA 2 mg/l	1/2 MS IAA 0.2–1.0 mg/l Kin 0.5–2.0 mg/l		Xin and Zhang (1987)
Apical and lateral buds	Kumura, Owairaka Red 888 Owairaka Red 907		MS NAA 1 mg/l	MS	Elliott (1969)
	White Star PI 315343		MS BA 1 mg/l (or IAA 1 mg/l Kin 1 mg/l)		Litz and Conover (1978)
	Tainung 57 Red Tip Tainan 15 Okinawa 100		MS IAA 1–10 mg/l (or IBA 1 mg/l) (or NAA 0.5 mg/l)	MS	Chen (1981)
	Centennial Jewel Redmar		MS BA 2 mg/l NAA 0.1 mg/l		Hwang et al. (1983)

Tissue	Cultivar				Reference
	White Star GaTG 3	MS 2,4-D 0.5–2.0 mg/l BA 0.3 mg/l	MS 2,4-D 0.5–2.0 mg/l CW 20% Kin 2 mg/l		Liu and Cantliffe (1984)
	CATIE PI 8458 CATIE PI 8463 CATIE PI 8467 CATIE PI 8473 CATIE PI 8485 CATIE PI 8491		MS 2,4-D 0.1–3.0 mg/l	1/2 MS	Jarret et al. (1984)
	Jewel		MS IAA 1 mg/l		Templeton-Somers and Collins (1986b)
Stem	Tainung New 31	MS 2,4-D 2 mg/l IAA 2 mg/l Kin 2 mg/l	MS AD 20 mg/l Kin 0.5 mg/l (or IAA 10 mg/l BA 1 mg/l) (or ABA 5–10 mg/l Kin 0.5 mg/l)	1/2 MS AS 7.5 mg/l Kin 0.5 mg/l NAA 0.1 mg/l	Tsay et al. (1982)
	Jewel Caromex	MS BA 10 mg/l NAA 1 mg/l	MS BA 0.1 mg/l NAA 1 mg/l		Carswell and Locy (1984)
	White Star	MS	MS		Liu and Cantliffe (1984)
	GaTG 3	2,4-D 1 mg/l	2,4-D 2.0 mg/l CW 20% Kin 2 mg/l		
	Kokei 14	MS 2,4-D 1 mg/l ABA 10–20 mg/l	MS Kin 1–5 mg/l (or BA 5 mg/l)	1/2 MS GA$_3$ 1 mg/l NAA 0.1 mg/l	Murata and Miyaji (1986)

Table 7.1 *Continued*

Explant	Clone	Medium for			Reference
		Callus	*Organogenesis*	*Regeneration*	
	331 12-17 Qinnanhong	MS BA 0.01–0.1 mg/l NAA 2 mg/l	1/2 MS IAA 0.2–1.0 mg/l Kin 0.5–2.0 mg/l		Xin and Zhang (1987)
Anther	JP	Miller 2,4-D 1–2 mg/l Kin 2 mg/l	1/2 White Kin 1 mg/l YE 4 mg/l	Bonner Kin 0.01 mg/l	Kobayashi and Shikata (1975)
		MS 2,4-D 1 mg/l AD 10–20 mg/l	MS		Sehgal (1978)
	Tainung Hsiñ 31	MS 2,4-D 2 mg/l IAA 2 mg/l Kin 2 mg/l	MS ABA 0.1–1.0 mg/l	MS IAA 1.0 mg/l Kin 4 mg/l	Tsay and Tseng (1979)
	Tainan 15 Golden Tainung New 31	MS 2,4-D 0 or 2 mg/l IAA 2 mg/l Kin 2 mg/l	MS AD 20 mg/l (or ABA 1–5 mg/l) Kin 0.5 mg/l	1/2 MS AS 7.5 mg/l Kin 0.5 mg/l NAA 0.1 mg/l	Tsay *et al.* (1982)
Storage root	Yellow Jersey Jersey Orange Centennial	White IAA 1 mg/l 2,4-D 1 mg/l (or AS 4–40 mg/l) (or Kin 0.5 mg/l)	White AD 4 mg/l CW 5% GA_3 1 mg/l IAA or 2,4-D 0.01–0.5 mg/l Kin 0.5 mg/l		Gunckel *et al.* (1972)

Cultivar	Treatment 1	Treatment 2	Treatment 3	Reference
Chugoku 25 Norin 1	White 2,4-D 0.01 mg/l	White ABA 0–1.0 mg/l NAA 0.001–0.005 mg/l		Yamaguchi and Nakajima (1972)
	White 2,4-D 0.1–1.0 mg/l (or NAA 10–20 mg/l)	White Kin 1–10 mg/l NAA 1 mg/l	White 2,4-D 0.01–0.1 mg/l ABA 0.01–1.00 mg/l Kin 0.001 mg/l	Yamaguchi (1978)
Tainung Hsin 31 Bud Mutant	MS 2,4-D 2 mg/l IAA 2 mg/l Kin 2 mg/l	MS ABA 5–10 mg/l Kin 0.5 mg/l	1/2 MS AS 7.5 mg/l Kin 0.5 mg/l NAA 0.1 mg/l	Tsay et al. (1982)
Centennial	MS BA 2 mg/l NAA 0.1 mg/l	MS BA 2 mg/l NAA 0.1 mg/l		Hwang et al. (1983)
Jewel Caromex	MS BA 10 mg/l NAA 1 mg/l	MS BA 0.1 mg/l NAA 1 mg/l		Carswell and Locy (1984)
White Star GaTG 3	MS 2,4-D 1 mg/l	MS 2,4-D 2.0 mg/l CW 20% Kin 2 mg/l		Liu and Cantliffe (1984)

Abbreviations: ABA, abscisic acid: AD, adenine; AS, adenine sulphate: Alar (B-995, SADH), succinic acid-2.2-dimethyl hydrazide; BA, benzylamino purine; CCC (Cycocel), (2-chloroethyl)-trimethylammonium chloride or chlorocholine chloride; CW, coconut water; GA$_3$, gibberellic acid; KS, Kao and Michayluk basal medium; IAA, indole-3-acetic acid; IBA, indole-3-butyric acid: Kin, kinetin; MS, Murashige and Skoog basal medium; NAA, naphthalene acetic acid; YE, yeast extract; 2,4-D, 2,4-dichlorophenoxy acetic acid.

131

Yamaguchi and Nakajima, 1972). Adventitious shoot and root formation may occur simultaneously or sequentially; mostly shoots arise in association with *de novo* formed roots on the callus (Carswell and Locy, 1984).

In some cases, the addition of plant hormone(s) to the nutrient medium also induces calli to give rise to somatic embryods (embryoids) through somatic embryogenesis (Tsay and Tseng, 1979; Hwang *et al.*, 1980; Jarret *et al.*, 1984; Liu and Cantliffe, 1984). This process is much less common than organogenesis. Embryoids are formed by repeated mitotic divisions of individual cells, generally located on the surface of callus tissues. These asexually produced embryo-like structures are bipolar with a root–shoot axis, lack a vascular connection to the maternal tissue from which they arise, and function like a zygotic embryo in giving rise to a plant. Much of the experimental evidence on sweet potato indicates that differentiation of embryogenic cells takes place in the presence of an auxin, particularly 2,4-D (Jarret *et al.*, 1984; Liu and Cantliffe, 1984), but that for the further development of embryoids 2,4-D is not necessary or inhibitory. These embryoids are usually transferred to hormone-free media in order to let them develop into plantlets.

An unlimited number of plantlets can theoretically be rapidly regenerated by either organogenesis or somatic embryogenesis of the callus derived from the explanted tissue; where applicable, this is often the fastest method of plantlet multiplication. However, its general applicability is restricted because specific conditions for successful plantlet regeneration depend on specific explants and clones after trial and error. In general, there is significant variation among families, progeny within families, and different tissue-culture protocols. A reproducible method such as hormonal concentrations and/or combinations for producing plantlets with high frequency of uniformity from callus is still not available.

There are other drawbacks to this method. The most serious objection to the use of callus cultures for plantlet multiplication is the genetic instability of their cells (Scowcroft, 1984). Despite the homogeneity of the initial explant and the possibility of establishing and continuing the true homogeneous nature of the initial explant on well-defined media, much variation can and does occur, especially on callus and subsequent subcultures. Regenerated plantlets are likely to have undergone genetic aberrations during the callus stage (Templeton-Somers and Collins, 1986b). This has been further demonstrated in callus itself (Oba and Uritani, 1979), regenerated root systems (Sihachakr and Osse, 1981), regenerated plantlets from explanted anthers (Tsay and Tseng, 1979), lateral buds (Litz and Conover, 1978; Jarret *et al.*, 1984), and storage-root tissue (Gunckel *et al.*, 1972; Yamaguchi and Nakajima, 1974). In a clonally propagated sweet potato these types of genetic changes would be unacceptable to the germplasm collector as well as to the grower. Another disadvantage of the multiplication method involving a callus phase is that the initial plantlet regeneration capacity of the callus may decline with passage of time and eventually be lost. Thus, plantlet regeneration from explants which readily produce calli would be an ineffective approach for multiplication and conservation of sweet potato germplasm.

On the contrary, the major advantage of cloning sweet potatoes through apical- and/or lateral-bud culture is the reduced or total lack of callus formation. The explant used to start cultures of this kind is a lateral or main shoot apex which may be up to 20 mm in length. Incorporation of auxins and/or cytokinins into the growth medium are able to promote the precocious development of otherwise inhibited meristems in apical and lateral buds (Table 7.1). Cultured apical and lateral buds usually develop into a single axis rather than clusters of miniature shoots (Elliott, 1969; Litz and Conover, 1978). Root formation on the developed shoots presents no serious difficulty and may happen with very little callus formation, thereby reducing the occurrence of genetic abnormalities.

Shoot-tip culture and/or lateral-bud culture is the preferred term to describe culture of apical and/or lateral buds, mainly to distinguish them from meristem culture. Much smaller explants are used to initiate the latter, and the aim is usually to produce a single virus-free plantlet from each explant.

7.3 MERISTEM CULTURE

Sweet potatoes as a vegetatively propagated crop are systematically infected with one or more pathogens. Viruses disseminate with the propagule through successive generations, so that the entire crop can become infected. Pathogens attack does not always lead to the death of the plant. Many viruses may not even show visible symptoms. However, the presence of viruses in the plants can not only reduce crop yield and/or quality of sweet potato (Chung et al., 1981; Chiu et al., 1982; Luo, 1988a,b), but can also seriously limit international germplasm exchange. Disease-free clones have been regenerated from disease-infected sweet potatoes through in vitro culture of the extreme tip of the shoot. The technique can be applied to all known infectious agents, but is especially valuable in eliminating viruses and similar organisms from this vegetatively propagated crop.

The growing point of a shoot consists of an apical dome of cells capable of active cell division (the apical meristem). Beneath the meristem there are ridges of progressively increasing size, which represent newly formed leaf initials. Because virus movement is considerably slower through the symplasm than through the vascular system, and vascular differentiation occurs away from the shoot apical meristem, virus concentration decreases acropetally, and in the growing point the virus titre may be nil. In addition, mitosis in the meristem cells competes with virus multiplication. Isolation and culture of meristem tips, therefore, may eliminate viruses present in the plant. Furthermore, *in vitro* conditions with the presence of hormones may also help to inactivate the virus. If successful, culture ultimately gives rise to small shoots. With appropriate treatments, the regenerated shoots can be rooted to produce disease-free plantlets.

Meristem-culture techniques involve the surface sterilization of shoot apex and/or lateral buds, the dissection of shoot apical meristematic domes plus one, two, or three rudimentary leaf initials, and then culturing them on a nutrient medium in which differentiation and complete plantlet development take place. Virus-free sweet potatoes derived from the use of meristem culture are presently being grown widely (Nielsen, 1960; Mori, 1971; Over de Linden and Elliott, 1971; Alconero *et al.*, 1975; Liao and Chung, 1979; Frison and Ng, 1981). The technique has been used to eradicate one or more sweet potato viruses such as feathery mottle virus (Mori, 1971; Liao *et al.*, 1982), internal cork virus (Nielsen, 1960; Mori, 1971), sweet potato mild mottle virus (Hollings *et al.*, 1976), sweet potato virus D complex (Frison and Ng, 1981), sweet potato latent virus (Chung *et al.*, 1986), sweet potato yellow dwarf virus (Green and Luo, 1989), and unidentified viruses (Over de Linden and Elliott, 1971; Alconero *et al.*, 1975; Luo, 1988a,b). The method also eliminates mycoplasm-like organisms from witches' broom-infected sweet potatoes (Green *et al.*, 1989). The technique is routinely used at AVRDC, CIP and IITA, and in some national programmes to eliminate virus diseases from sweet potato (Dodds and Ng, 1988; Green, 1988; IBPGR, 1988; Moyer, 1988).

Methods and growing conditions of meristem culture of sweet potato for eradicating viruses are described in detail by Love *et al.* (1987). Because all stock materials are not necessarily pathogen-free, the method generally includes identifying the pathogens, especially viruses, in the stock material. Subsequent procedures include applying techniques for eliminating diseases from stock materials, testing regenerated plantlets for freedom from pathogens, and propagating healthy plants under conditions that prevent reinfection. The IBPGR has stressed to its collaborators worldwide the need for this type of work and it is built into the *in vitro* gene-bank design (Withers and Williams, 1985; Withers, 1986; IBPGR, 1988).

Success in virus elimination using meristem culture depends on virus type, host plant, source of meristems, and size of the meristem used for culture. The proportion surviving in meristem culture is usually greater when explants have been taken from the growing points of plants, rather than from lateral buds. This may be because apical meristems are naturally in active growth, while meristems in lateral buds are dormant. The size of the meristematic dome also determines their ability to survive on nutrient medium even at optimal conditions. The larger the meristem, the greater are the chances of survival and plantlet regeneration. The small size of the meristem may have adverse effects in forming either callus or root only. In the present context, however, the survival and regeneration of the meristems cannot be treated independently of the efficiency with which virus elimination is achieved, which is inversely related to the size of the meristem. Thus, meristems should be small enough to eradicate viruses and large enough to be able to develop into a complete plantlet. Besides the size of the meristem, the presence of leaf primordia improves the ability of the meristems to form plantlets.

Stock materials are usually treated with fungicides and insecticides (Dodds and Ng, 1988; Green, 1988) before stem-tip and/or lateral-bud cutting, and meristem excision. Explants are initially taken from stem tips and axillary buds of pathogen-tested, field-grown plants, or more commonly from sprouts removed from pathogen-tested storage roots. The latter are favoured as they have a better survival rate and less disease infection than field-grown plants. Rapidly growing sprouts are preferred. Stem tips about 2–3 cm in length are removed from sprouts. They are surface-sterilized for 10–15 min in a solution of sodium hypochlorite (0.5%) and the surfactant Tween 20 (0.02%). After rinsing well in sterile water, they are taken to sterile conditions, usually a laminar-flow cabinet. Under a stereo microscope, their overlapping leaf primordia are removed by a set of needles and blades mounted on separate handles. Meristems are then excised and transferred to test tubes containing 10 ml of medium for culturing.

For virus-elimination purposes, the procedure of meristem culture usually involves excision of shoot-tip explants (0.4–0.8 mm) comprising the apical meristem (0.1–0.2 mm) and at least one primordial leaf. In some clones it is difficult to excise a meristem which is small enough to avoid virus infection but which is still capable of regeneration to form a plantlet. It is, however, possible to obtain virus-free plantlets from such infected clones by culturing relatively larger meristems in combination with heat treatment before meristem excision. The heat treatment is usually carried out on the whole storage root with sprouts by growing it in a controlled-temperature cabinet at 30–40°C for a period of 4–6 weeks (Liao and Chung, 1979; Chiu et al., 1982; Green, 1988). High temperatures probably do not eliminate most viruses, but are less favourable for viral replication and translocation within the plant. Unfortunately, the optimum time for minimal virus infection may not always coincide with the period when meristem culture is most readily initiated, and excessive exposure to high temperature may adversely affect plant tissues.

The MS medium containing 3–5% sucrose has been found useful in culturing meristems from a wide range of sweet potato clones (Table 7.2). Because of the convenience in handling, 0.7–1.0% agar medium is generally preferred. In most cases, the meristems are cultured under conditions of 1500–5000 lux on a light period of 12–16 h at 28–30°C. Various growth regulators accelerate the growth and development of bud and root initiation, and eventually plantlet regeneration. Based on published data (Table 7.2), the optimum concentration and combination of growth regulators for shoot and root formation may be genotype-dependent. It is possible that these clonal differences are due to different endogenous concentrations of plant hormones. In most cases, IAA or NAA 1 mg/l in combination with kinetin 1 mg/l, or IBA 10 mg/l alone promote root differentiation and shoot growth of the cultured meristems with minimal callus formation (Kuo *et al.*, 1985; Scaramuzzi, 1986; Love *et al.*, 1987). The presence of BA (10 mg/l) in the medium often causes callus formation; however, activated charcoal is used to suppress excess callus formation in favour of regenerated shoots (Litz and Conover, 1978). Also, ethylene at 9 ppm is known to inhibit callus formation on injured surfaces of explants (Chalutz and DeVay, 1969). The normal process leading to plant regeneration is swelling of the meristem, callus formation in some instances, shoot development, and finally root formation. The whole process from the initial meristem culture to a fully grown plantlet with four to five nodes may take 2–4 months.

7.4 *IN VITRO* MULTIPLICATION

Once plantlets are established from meristem culture, the rapid multiplication of plantlets can be accomplished by *in vitro* culture of single nodal cuttings from a regenerated plantlet under *in vitro* conditions (Litz and Conover, 1978). When they are placed onto an appropriate culture medium, the axillary bud of the meristem-derived plantlet is induced to grow and initiate roots, and results in the formation of the new plantlet (Sihachakr, 1982). The basic advantages of this *in vitro* propagation lie mainly in the rapidity with which plant multiplication can be achieved and the number of plantlets that can be produced in a relatively short period of time, with conservation of space, and often at a lower cost.

The hormone/nutrient conditions of the medium play a role in promoting the meristem in the axillary bud to grow and differentiate a root system. Media for single nodal propagation of sweet potato have been compared (Love *et al.*, 1987); the nodal cuttings grow well in the same medium as was used for the original meristem culture. However, care must be taken that the culture conditions do not allow callus formation before regeneration of plantlets. Regeneration of plantlets from nodal cuttings usually takes place

Table 7.2 Summary of *in vitro* plantlet regeneration of sweet potato from meristems (< 0.5 mm)

Source	Medium for	
	Organogenesis	Regeneration
Nielsen (1960)	Nielsen	
Elliott (1969)	MS	MS
	NAA 1 mg/l	
Mori (1971)	1/2 Knop	
Alconero *et al.* (1975)	MS	
	IAA 0.2–2.0 mg/l	
	Kin 0.5–5.0 mg/l	
Chen (1978,1980)	MS	
	AS 7.5 mg/l	
	Kin 0.5–1.0 mg/l	
	NAA 0.1 mg/l	
Liao and Chung (1979)	MS	MS
	BA 4–8 mg/l	Kin 2 mg/l
	IAA 1–2 mg/l	
Frison and Ng (1981)	MS	
	BA 0.5 mg/l	
	IAA 0.2 mg/l	
Kuo *et al.* (1985)	MS	
	IAA 1 mg/l	
	Kin 1 mg/l	
Rey and Mroginski (1985)	MS	
	GA$_3$ 1 mg/l	
	Kin 0.1 mg/l	
	NAA 0.1 mg/l	
Scaramuzzi (1986)	MS	MS
	Kin 1 mg/l	IBA 1 mg/l
	NAA or IAA 1 mg/l	
Love *et al.* (1987)	MS	MS
	BA 0.3 mg/l	
	NAA 0.03 mg/l	
Otsuki *et al.* (1987)	MS	
	BA 1 mg/l	
	NAA 0.01 mg/l	
Luo (1988a,b)	MS	MS
	BA 0.5 mg/l	IAA 0.1 mg/l
	GA$_3$ 0.5–1.0 mg/l	
	IAA 0.01 mg/l	

For abbreviations see Table 7.1.

under conditions of 16 h of 3000–5000 lux light at 25–28°C. Under these conditions *in vitro* propagation rates are fast, and a single nodal cutting can grow into a fully grown plantlet 5–7 cm in height with four or five nodes and a healthy, well-developed root system within 4–6 weeks.

After the regenerated plantlets have been multiplied and established, stringent virus-indexing procedures must be employed to determine the virus-free status of plantlets regenerated from infected source materials. Methods of virus testing depend on the virus concerned and the facilities available. Electron microscopic examination of leaf and sap material, serum-specific electron microscopy and serological tests such as enzyme-linked immunosorbant assay (ELISA) have provided methods of greater sensitivity. However, these sensitive and very rapid indexing methods are insufficient because they can only be applied to one or two of the virus complexes. For this reason, most commonly used indexing techniques for regenerated plants usually involve the long but very sensitive procedure of sap transmission through grafting to indicator plants, commonly *Ipomoea setosa* and sensitive *I. batatas* clones. Schemes for disease elimination in sweet potato are described elsewhere (Love *et al.*, 1987; Dodds and Ng, 1988; Green, 1988). Unfortunately, the virus-indexing procedures described here cannot guarantee complete freedom from all viruses. This is due to specificity and limited sensitivity of employed virus tests; efforts to overcome these problems are still in progress (Moyer, 1988).

In the described procedure, some plantlets from nodal cuttings are taken out of culture and their tissues are tested for viruses by the ELISA test (Green, 1988). The remaining plantlets derived from the same meristem, whose sister plantlets have been tested and shown to be free from viruses, can be multiplied under *in vitro* conditions by repeating the nodal-propagation system as often as necessary. Virus-tested plantlets could be multiplied at a rate of four plantlets per month by the *in vitro* single-node cutting methods (Frison and Ng, 1981; Kuo *et al.*, 1985). This means that each month a bud could provide a plantlet from which an average of four buds could be recultured. At that rate, using a geometric progression, thousands of plantlets could be produced in 12 months. This method of multiplication minimizes the risk of virus recontamination, and the virus-tested plantlets can serve as a source of stock plant material for germplasm repositories. The method also facilitates international transport and exchange of germplasm, has the potential to upgrade quarantine services, and is increasingly used to distribute disease-tested sweet potatoes to plant breeders worldwide.

The plantlets derived from nodal cuttings can easily be transferred to potting mixtures of soil, sand, peat, and/or vermiculite. They are set under non-sterile but insect-proof conditions, usually initially under high humidity and low light conditions. Further technical improvements are necessary to minimize the loss of many plantlets at this final stage of clonal propagation. After being grown out into full plants, they are usually further tested for the presence of virus by grafting a node with a fully expanded leaf to indicator plants (Love *et al.*, 1987). Sister plants of the same stock plantlet can be maintained and further multiplied in insect-proof glass or gauze houses to produce virus-tested stock materials for field production.

In vitro exchange of sweet potato germplasm has been used for a number of years by national and international institutes to distribute germplasm efficiently and securely. Practical procedures for culture package, handling before and after transit, certification and *in vitro* transfer are well established and described elsewhere (Love *et al.*, 1987; IBPGR, 1988).

7.5 *IN VITRO* CONSERVATION

The use of *in vitro* techniques in maintaining germplasm of vegetatively propagated sweet potatoes has the advantages of preserving healthy plant materials in a small space, ease and rapid multiplication for international exchange, and cost reduction. There are certain prerequisites for the successful application of *in vitro* gene banks, such as efficient methods to introduce tissues aseptically into cultures and to regenerate plants from culture, and the development of storage conditions that offer a high level of survival, stability, and reproducibility in the culture. For these reasons, callus cultures are generally considered to be unsuitable for germplasm storage. The method that appears to be most widely applicable is apical- and lateral-bud culture.

Culture maintenance under the standard *in vitro* system of bud culture, due to its high multiplication rates, requires serial subcultures. This results in increased expense, risk of genetic instability and microbial contamination, and loss through deterioration, human error or equipment failure. Thus, the maintenance of bud cultures in continued growth at a normal rate is considered to be an undesirable approach in germplasm conservation; the practical approach is to reduce the frequency of subcultures to a bare minimum.

For short-term storage, the maintenance of bud cultures under growth-limiting conditions offers the possibility of reducing the requirement of subcultures. Prominent among technical approaches are the use of growth-inhibiting chemicals (natural and synthetic hormones and osmotically active agents) and moderate reduction in the temperature at which cultures are maintained. Slowing culture rates usually extends subculture intervals to as much as one year or more. These methods have the advantages that the stored material is readily available for use and that the stocks are replenished as the cultures grow.

The optimal growth temperature for plantlet development appears to be between 28°C and 30°C. However, the growth rate of nodal cuttings in MS medium without growth hormones can be restricted by decreasing the incubation temperature to about 16–20°C for storage purposes (Ng and Hahn, 1985; Love et al., 1987; Xin, 1987; Luo, 1988a,b). Combinations of reduced temperatures and low light conditions of 1000 lux are considered to be appropriate for short-term storage, about one year, of sweet potato germplasm. To achieve reduced temperature simply requires a culture room or incubator with appropriate temperature control. It should be noted, however, that temperatures lower than 12°C may be lethal to in vitro cultured sweet potato materials.

Addition to the culture medium of osmotic active agents, which have inhibitory but non-toxic effects on cell growth, may also be effective in reducing bud growth by acting directly to induce bud dormancy, or to reduce cell metabolism and division. One of the advantages of this method is that the modification of the culture medium requires no special equipment. High mannitol levels, approximately at 1.0–1.5% in the culture medium, can therefore be used to maintain sweet potato plantlets in a dormant condition for long periods, about 12 months, with a 70% survival rate (Love et al., 1987; Xin, 1987,1988). Mannitol is likely to preserve membrane integrity and prevent solute leakage. In some cases, high mannitol levels are given in combination with reduced temperature (16–20°C) and light intensity (1000–2000 lux). This is considered to be the most realistic and cost-effective way to maintain a large sweet potato germplasm collection (Dodds, 1988); addition of mannitol would make cells less susceptible to cold. However, a higher level of mannitol, approximately at 3% in the culture medium (Xin, 1988), may be effective in reducing growth at the normal temperature, thus allowing stored cultures to be maintained alongside those bud cultures for normal shoot development. Other osmotically active agents include sorbitol and high sucrose levels, about 6–8%. The preferred treatments and precise combination will vary from one laboratory to another, and according to the type of clone being stored.

Other modifications of the culture medium are also effective. The MS medium without auxins but with 7.5 mg/l of adenine sulphate is able to induce the bud in dormancy for more than 18 months without losing its totipotency (Chen, 1978). Cultures of buds at 16–20°C and under low light intensity at 1000 lux in the presence of the growth retardant Alar (B-995) (Love et al., 1987), or some other growth inhibitor or retardant, i.e. CCC 100–500 mg/l, ABA 1 mg/l or glyphosate 1 mg/l (Xin, 1987), provide other methods by which sweet potato germplasm can be preserved for about 12 months with at least 70% survival rates. Similar effects are noted in the MS medium incorporating 5–10 mg/l of kinetin (Xin, 1987,1988). The difficulty with medium modified with growth regulators is that different clones may react differently under these conditions.

Under suitable minimal-growth conditions, the use of polypropylene film or tin foil allows a useful compromise between gas exchange and moisture retention, since the evaporation of medium during long storage periods can cause water loss, yet sealing the vessel drastically reduces survival, due to prevention of gas exchange. Taking all favourable treatments, storage of sweet potato plantlets in slow growth for up to two years appears to be feasible (Ng and Hahn, 1985).

7.6 CELL AND PROTOPLAST CULTURE

In sweet potato, progress in the cryopreservation of cell and protoplast cultures is still limited by deficiencies in the regenerative capacity of cells and protoplasts in culture. With the exception of a few reports describing plantlet regeneration (Murata *et al.*, 1987a; Sihachakr and Ducreux, 1987a,b), the culture of cells and protoplasts has usually led to callus formation only. The potential of plant regeneration from cell suspensions, and the fact that protoplasts can give rise to plantlets, are promising for the future. At present, however, there are only a few reports of all cultures which are stable in the long term. Non-morphogenetic cell lines remain the only accessible material in which to study cryopreservation. Despite obvious limitations, they are suitable for initial experiments directed towards somaclonal variation and the introduction of foreign genes into cultured cells and/or protoplasts (Scowcroft, 1984).

In this section, reports on cell and protoplast culture of sweet potato are reviewed. They are a rudimentary step towards the development of efficient and reproductive cryopreservation for sweet potato. Obviously, future improvements for the purpose of long-term storage are needed.

7.6.1 Culture initiation

Establishment of a friable callus consisting of rapidly dividing cells seems to be a critical step in the cell culture of sweet potato; such a callus could be obtained either by selecting special explants or by using special media (Schwenk, 1981). In some instances the sloughing off of cells from the explanted tissue or organ may serve as a source of cell culture. The usual procedure is to establish a cell-suspension culture from a friable callus in a liquid medium that is physically agitated or rotated by some mechanical device. A high K/Ca ratio in the MS medium containing 2,4-D, NAA, kinetin, and thiamine-HCl appeared to be good for cell culture (Yoshida *et al.*, 1970). A degradation product of liquid cell culture, 2,3-diketogluconic acid, on the other hand, inhibited cell growth (Nakamura *et al.*, 1981).

Protein profiles and isoenzymic composition of cultured cells on liquid or solidified Gamborg's PRL-4 medium with casein amino acids were remarkably different from those of native plant tissue (Sasaki *et al.*, 1972). In contrast to concerns with genetic instability, the phenotypic variability observed among cultured cells has been used for the isolation of spontaneous or induced mutant cell lines (Scowcroft, 1984). In this connection, a new or modified amylase, an enzyme responsible for starch degradation, was produced by the liquid-cultured cells (Handley and Locy, 1984), whereas starch formation in cell culture was inhibited by GA 3 but not NAA, BA and ABA, suggesting that the biosynthesis of starch might be repressed by GA_3 (Sasaki and Kainuma, 1984). Furthermore, salt-resistant cells could be isolated by subculturing cell suspensions in MS medium supplemented with 1% NaCl; the acquired trait was stable for at least three generations (Salgado-Garciglia *et al.*, 1985).

7.6.2 Protoplast isolation and culture

Successful isolation of protoplasts from stem callus of sweet potato was first reported by Wu and Ma (1979). Protoplasts were released by enzymatic digestion of the middle lamellae with cellulase (2.0%) in 20 mM $CaCl_2.2H_2O$ and 0.8 M mannitol within 5 h. Protoplasts regenerated new cell walls after seven days in agar media containing 2,4-D 0.1 mg/l and kinetin, or NAA 0.3 mg/l and kinetin 0.1 mg/l. Cell division took place after ten days, and calli were observed after 40 days.

Since mesophyll tissue was somewhat recalcitrant to the enzyme digestion, petioles were used to isolate protoplasts (Bidney and Shepard, 1980). Protoplasts were induced to form callus, from which roots were produced on culture medium containing NAA 0.05 mg/l and BA 0.5 mg/l within three weeks. Similar results were obtained from petioles and stems by Sihachakr and Ducreux (1987a,b), but there was an influence of organs and genotypes on the development of protoplasts in culture. Protoplasts could also be isolated from high-anthocyanin-producing callus originated from storage root (Nishimaki and Nozue, 1985). The highest protoplast yield was obtained from 0.6 g of six-day-old callus incubated with 6 ml of enzyme (cellulase and pectolyase) solution. Optimum culture conditions were as follows: protoplasts ($1-3 \times 10^4$ cells/ml) were cultured in 2 ml of modified PRL-4-C medium ($CaCl_2.2H_2O$ 300 mg/l) supplemented with 0.5 M mannitol, 0.1 M sucrose, casein amino acids 1000 mg/l, 2,4-D 0.2 mg/l, kinetin 1 mg/l and 0.5% Bacto-Agar.

Successful plant regeneration from protoplasts of sweet potato has been demonstrated. Sihachakr and Ducreux (1987a,b) isolated protoplasts from petioles and stems and cultured them in KM8P medium supplemented with 2,4-D 0.2 mg/l, zeatin 0.5 mg/l and NAA 0.5 mg/l at a density of 3×10^4 protoplasts/ml at 5 ml per dish. Green compact calli with meristematic areas were later induced in the medium supplemented with zeatin 2 mg/l, and 5% of plantlet regeneration occurred when these calli were transferred onto the medium with 0.25 mg/l zeatin. Murata *et al.* (1987a) derived protoplasts from petioles. They were cultured in modified MS medium containing NH_4NO_3 200 mg/l, KNO_3 1000 mg/l, kinetin 0.1 mg/l, 2,4-D, 0.5 M mannitol and 1% sucrose. Shoot formation was induced from the callus on a medium supplemented with 1 mg/l kinetin, and only when preceded by culture on a medium containing 0.5 mg/l of 2,4-D and kinetin, and ABA 1 mg/l. Later, glutamic acid, asparagine and proline were found to enhance plantlet regeneration (Murata *et al.*, 1987b).

Furthermore, cell colonies were obtained through protoplast fusion between *I. batatas* and its wild related species, *I. triloba* (Kokubu and Sato, 1987). Protoplast fusion was carried out by the polyethylene glycol method using protoplasts isolated from green callus of *I. batatas* and white callus of *I. triloba* originated from petioles. The fusion-treated protoplasts cultured in agar medium produced cell colonies within a few days, but as yet no callus or shoot morphogenesis, or embryogenesis, has occurred.

7.7 GENETIC STABILITY OF CULTURES

Germplasm of clonally propagated crops is likely to contain more genetic variation than in species in which seed is the form of conservation. For sweet potato germplasm conservation it is important to know that each propagule is true to type. In the case of its *in vitro* conservation, high levels of phenotypic variability or abnormality and even the loss of totipotency might be observed in regenerated plantlets because of chromosomal aberrations, point mutations, and biochemical changes (Withers, 1986). Even small recessive genetic changes, which are not expressed phenotypically because of a high level of heterozygosity and ploidy, may accumulate from one generation to the next and eventually may affect uniformity. If a plantlet should regenerate from storage with a different characteristic to that possessed by the material from which it originated, then the validity of the storage technique must be re-examined.

The causes of genetic changes during tissue culture are still obscure. Populations of plantlets possessing a significant degree of genetic variation can, however, be produced through a high rate of multiplication and the use of inappropriate *in vitro* techniques. The auxin 2,4-D is often implicated in a high frequency of genetic abnormalities, and chelating agents and some micronutrient metals have been noted to induce chromosome breakage (Scowcroft, 1984).

Although maintaining sweet potato bud culture under growth-limiting conditions offers the possibility of reducing the requirement for frequent subculturing, genetic instability might arise due to the imposition of environmental stress on plantlets over a long period of storage. However, a lack of scientific data on the genetic stability of sweet potato cultures in slow growth – or even in normal growth – point to the conclusion that a clear statement on their incidence is not yet possible. Evidence from sweet potato and its related species shows that plantlets derived from the meristem culture and nodal propagation generally exhibit little or no genetic differences from their parents (M. Kobayashi, unpublished results). Despite this fact, for purposes of practical procedures incorporating an adequate margin of safety, a figure of up to 5% incidence of off-types per subculturing cycle is suggested (IBPGR, 1981). Therefore, it is still essential to monitor small changes in the horticultural characteristics such as plant type, storage root colour (skin and flesh), growth habit and other morphological characters of regenerated plants (Dodds and Ng, 1988). Furthermore, a number of biochemical methods such as soluble proteins and isoenzymes (peroxidase) are also currently used in sweet potato for detecting variation (Otsuki et al., 1987; Dodds, 1988). Although these biochemical methods appear to be effective for looking at variation in mature plants, no data are available on in vitro plantlets. Only when exact data are available, and it is known that there is indeed a satisfactory level of genetic stability, can the numbers of cultures held in storage be reduced. Moreover, it is important to stress that, in the main, in vitro methods should form an adjunct to other methods of conservation rather than be the sole method of conservation.

REFERENCES

Alconero, R., Santiago, A. G., Morales, F. and Rodriguez, F. (1975) Meristem-tip culture and virus indexing of sweet potatoes. *Phytopathology*, **65**, 767–773.

Anonymous (1982) *Sweet Potato Cultivation in China* (in Chinese), Shanghai Sci. Tech. Publ. Co., Shanghai, China.

Bidney, D. L. and Shepard, J. F. (1980) Colony development from sweet potato petiole protoplasts and mesophyll cells. *Plant Sci. Lett.*, **18**, 335–342.

Bouwkamp, J. C. (1984) *Sweet Potato Products: A Natural Resource for the Tropics*, CRC Press, Florida, USA.

Carswell, G. and Locy, R. D. (1984) Root and shoot initiation by leaf, stem, and storage root explants of sweet potato. *Plant Cell Tissue and Organ Culture*, **3**, 229–236.

Chalutz, E. and DeVay, J. E. (1969) Production of ethylene *in vitro* and *in vivo* by *Ceratocystis fimbriata* in relation to disease development. *Phytopathology*, **59**, 750–755.

Chen, S. J. (1978) The effect of cytokinin and auxin on the shoot tip culture of *Ipomoea batatas*. *J. Sci. Eng. Chunghsing Univ. (Taiwan)*, **15**, 145–156.

Chen, S. J. (1980) The shoot tip culture of several varieties of *Ipomoea batatas*. *J. Sci. Eng. Chunghsing Univ. (Taiwan)*, **57**, 203–208.

Chen, S. J. (1981) Regeneration of sweet potato from different parts of shoot tip *in vitro*. *J. Sci. Eng. Chunghsing Univ. (Taiwan)*, **58**, 107–118.

Chiu, R. J., Liao, C. H. and Chung, M. C. (1982) in *Sweet Potato. Proc. First Int. Symp.*, (ed. R. C. Villareal and T. D. Griggs), AVRDC, Taiwan, pp. 169–177.

Chung, M. L., Liao, C. H. and Li, C. (1981) Effect of virus infection on the yield and quality of sweet potatoes. *Plant Protection Bull. (Taiwan)*, **23**, 137–141.

Chung, M. L., Hsu, Y. H., Chiu, R. J. and Che, M. J. (1986) in *Virus Diseases of Horticultural Crops in the Tropics and Subtropics*, Food Fert. Tech. Ctr, Taiwan, pp. 84–90.

De la Puente, F. (1988) in *Exploration, Maintenance, and Utilization of Sweet Potato Genetic Resources*, CIP, Peru, pp. 75–100.

Dodds, J. H. (1988) in *Exploration, Maintenance, and Utilization of Sweet Potato Genetic Resources*, CIP, Peru, pp. 185–192.

Dodds, J. H. and Ng, S. Y. C. (1988) in *Exploration, Maintenance, and Utilization of Sweet Potato Genetic Resources*, CIP, Peru, pp. 323–329.

Elliott, R. F. (1969) Growth of excised meristem-tips of Kumara, *Ipomoea batatas* (Linn.) Poir. in axenic culture. *N. Z. J. Bot.*, **7**, 158–166.

Frison, E. A. and Ng, S. Y. (1981) Elimination of sweet potato virus disease agents by meristem tip culture. *Trop. Pest Mgt*, **27**, 452–454.

Green, S. K. (1988) in *Exploration, Maintenance, and Utilization of Sweet Potato Genetic Resources*, CIP, Peru, pp. 311–317.

Green, S. K. and Luo, C. Y. (1989) Elimination of sweet potato yellow dwarf virus (SPYDV) by meristem tip culture and heat treatment. *J. Plant Dis. Protection*, **96**, 464–469.

Green, S. K., Luo, C. Y. and Lee, D. R. (1989) Elimination of mycoplasm-like organisms for witches' broom infected sweet potato. *J. Phytopathol.*, **126**, 204–212.

Gunckel, J. E., Sharp, W. R., Williams, B. W., West, W. C. and Drinkwater, W. O. (1972) Root and shoot initiation in sweet potato explants as related to polarity and nutrient media variations. *Bot. Gaz.*, **133**, 254–262.

Handley, L. W. and Locy, R. D. (1984) Amylase secretion by sweet potato,

Ipomoea batatas (L.) Lam., cell cultures grown on various carbon sources. *Plant Cell Tissue and Organ Culture*, **3**, 237–245.

Henderson, J. M. M., Phills, B. R. and Whatley, B. T. (1983) in *Handbook of Plant Cell Culture 2* (ed. W. R. Sharp, D. A. Evans, P. V. Ammirato and Y. Yamada), Macmillan, New York, USA, pp. 302–326.

Hollings, M., Stone, O. M. and Bock, K. R. (1976) Purification and properties of sweet potato mild mottle, a whitefly borne virus from sweet potato (*Ipomoea batatas*) in East Africa. *Ann. Appl. Biol.*, **82**, 511–528.

Horton, D. E. (1988) in *Exploration, Maintenance, and Utilization of Sweet Potato Genetic Resources*, CIP, Peru, pp. 17–25.

Hwang, L. S., Skirvin, R. M., Casayo, J. and Bouwkamp, J. (1980) Embryoid formation from sweet potato root disc *in vitro*. *HortScience*, **15**(3), 415 (abstract).

Hwang, L. S., Skirvin, R. M., Casayo, J. and Bouwkamp, J. (1983) Adventitious shoot formation from sections of sweet potato grown *in vitro*. *Sci. Hort.*, **20**, 119–129.

IBPGR (1988) *Conservation and Movement of Vegetatively Propagated Germplasm:* In vitro *Culture and Disease Aspects*, IBPGR, Rome, Italy.

Jarret, R. L., Salazar, S. and Fernandez, R. Z. (1984) Somatic embryogenesis in sweet potato. *HortScience*, **19**, 397–398.

Kobayashi, M. and Shikata, S. (1975) Anther culture and development of plantlets in sweet potato. *Bul. Chugoku Natl Agri. Exp. Sta.*, **A24**, 109–124.

Kokubu, T. and Sato, M. (1987) Protoplast fusion of sweet potato and its related species. *Jpn. Soc. Breeding*, **72**, 56–57 (abstract).

Kuo, C. G., Shen, B. J., Shen, M. J., Green, S. K. and Lee, D. R. (1985) Virus-free sweet potato roots derived from meristem-tips and leaf-cuttings. *Sci. Hort.*, **26**, 231–240.

Latta, R. (1971) Preservation of suspension cultures of plant cells by freezing. *Can. J. Bot.*, **49**, 1253–1254.

Liao, C. H. and Chung, M. L. (1979) Shoot tip culture and virus indexing in sweet potato. *J. Agri. Res. China*, **28**, 139–144.

Liao, C. H., Tsay, H. S. and Lu, Y. C. (1982) Studies on eradication of SPV-A and SPV-N viruses from infected sweet potato. *J. Agri. Res. China*, **31**, 239–245.

Litz, R. E. and Conover, R. A. (1978) *In vitro* propagation of sweet potato. *HortScience*, **13**, 659–660.

Liu, J. R. and Cantliffe, D. J. (1984) Somatic embryogenesis and plant regeneration in tissue cultures of sweet potato (*Ipomoea batatas* Poir.). *Plant Cell Rpt*, **3**, 112–115.

Love, S. L., Rhodes, B. B. and Moyer, J. W. (1987) *Meristem Culture and Virus Indexing of Sweet Potatoes*, IBPGR, Rome, Italy.

References

Luo, H. Y. (1988a) Virus indexing and source preservation of sweet potato. *Guizhou J. Agri. Sci.*, **2**, 6–11.

Luo, H. Y. (1988b) in *Genetic Manipulation in Crops*, (ed. International Rice Research Institute), Cassell Tycooly, pp. 390–392.

Mori, K. (1971) Production of virus-free plants by means of meristem culture. *Jpn. Agri. Res. Q.*, **6**, 1–7.

Moyer, J. W. (1988) in *Exploration, Maintenance, and Utilization of Sweet Potato Genetic Resources*, CIP, Peru, pp. 303–310.

Murata, T. and Miyaji, Y. (1986) Regeneration of plants from stem callus of sweet potato. *Jpn. Soc. Breeding*, **69**, 24–25 (abstract).

Murata, T., Hoshino, K. and Miyaji, Y. (1987a) Callus formation and plant regeneration from petiole protoplast of sweet potato, *Ipomoea batatas* (L.) Lam. *Jpn. J. Breeding*, **37**, 291–298.

Murata, T., Hoshino, K., Fukoka, H. and Miyaji, Y. (1987b) Effects of several amino acids on plant regeneration from protoplast of sweet potato. *Jpn. Soc. Breeding*, **72**, 88–89 (abstract).

Nakamura, Y., Ikeda, Y., Ebihara, N. and Tabuchi, K. (1981) Inhibition of growth of callus cells (from carrot, tobacco, soybean, sweet potatoes) by degradation products of dehydroascorbic acid. *Agri. Biol. Chem.*, **45**, 759–760.

Ng, S. Y. and Hahn, S. K. (1985) in *Biotechnology in International Agriculture Research*, International Rice Research Institute, Philippines, pp. 29–40.

Nielsen, L. W. (1960) Elimination of the internal cork virus by culturing apical meristems of infected sweet potatoes. *Phytopathology*, **50**, 840–841.

Nishimaki, T. and Nozue, M. (1985) Isolation and culture of protoplasts from high anthocyanin-producing callus of sweet potato. *Plant Cell Rpt*, **4**, 248–251.

Oba, K. and Uritani, I. (1979) Biosynthesis of furano-terpenes by sweet potato cell culture. *Plant Cell Physiol.*, **20**, 819–826.

Otsuki, Y., Yoshida, Y., Abe, J., Tarumoto, I., Ishikawa, H. and Kato, S. (1987) Genetic variation occurring *in vitro* culture of rice and sweet potato. *Agri. Res. Ctr Annu. Rpt Lab.*, **I**, 1–10.

Over de Linden, A. J. and Elliott, R. F. (1971) Virus infection in *Ipomoea batatas* and a method for its elimination. *N. Z. J. Agri. Res.*, **14**, 720–724.

Rey, H. Y. and Mroginski, L. A. (1985) Efecto del ácido giberélico en la regeneración de plantas de batata (*Ipomoea batatas*) por cultivo *in vitro* de meristemas. *Phyton*, **45**, 123–127.

Salgado-Garciglia, R., Lopez-Gutierrez, F. and Ochoa-Alejo, N. (1985) NaCl-resistant variant cells isolated from sweet potato cell suspensions. *Plant Cell Tissue and Organ Culture*, **5**, 3–12.

Sasaki, T. and Kainuma, K. (1984) Control of starch and exocellular polysaccharides biosynthesis by gibberellic acid with cells of sweet potato cultured *in vitro*. *Plant Cell Rept*, **3**, 23–26.

Sasaki, T., Tadokoro, K. and Suzuki, S. (1972) The distribution of glucose phosphate isomerase isoenzymes in sweet potato and its tissue culture. *Biochem. J.*, **129**, 789–791.

Scaramuzzi, F. (1986) in *Biotechnology in Agriculture and Forestry* (ed. Y. P. S. Bajaj), Springer-Verlag, Berlin, Heidelberg, pp. 455–461.

Schwenk, F. W. (1981) Callus formation from mechanically isolated mesophyll cells of soybean and sweet potato. *Plant Sci. Lett.*, **23**, 147–151.

Scowcroft, W. R. (1984) *Genetic Variability in Tissue Culture: Impact on Germplasm Conservation and Utilization*, IBPGR, Rome, Italy.

Sehgal, C. B. (1978) Regeneration of plants from anther cultures of sweet potato (*Ipomoea batatas* Poir.). *Z. Planzenphysiol.*, **88**, 349–352.

Sihachakr, D. (1982) Premiers resultats concernant la multiplication vegetative *in vitro* de la patate douce (*Ipomoea batatas* L. Lam., Convolvulaceae). *Agron. Trop.*, **37**, 142–151.

Sihachakr, D. and Ducreux, G. (1987a) Plant regeneration from protoplast culture of sweet potato (*Ipomoea batatas* Lam.). *Plant Cell Rpt*, **6**, 326–328.

Sihachakr, D. and Ducreux, G. (1987b) Isolement et culture de protoplastes à partir de petioles et de tiges de deux variétés de patate douce (*Ipomoea batatas*). *Can. J. Bot.*, **65**, 192–197.

Sihachakr, D. and Osse, A. (1981) Analysis of root polymorphism observed in several explants of sweet potatoes (*Ipomoea batatas*) in *in vitro* culture. *Phytomorphology*, **31**, 112–121.

Suga, R. and Irikura, Y. (1988) Regeneration of plants from leaves of sweet potato and its related wild species. *Jpn. Soc. Breeding*, **73**, 56–57 (abstract).

Takagi, H. (1988) in *Exploration, Maintenance, and Utilization of Sweet Potato Genetic Resources*, CIP, Peru, pp. 147–157.

Templeton-Somers, K. M. and Collins, W. W. (1986a) Heritability of regeneration in tissue cultures of sweet potato (*Ipomoea batatas* L.). *Theor. Appl. Genet.*, **71**, 835–841.

Templeton-Somers, K. M. and Collins, W. W. (1986b) Field performance and clonal variability in sweet potatoes propagated *in vitro*. *J. Am. Soc. Hort. Sci.*, **111**, 689–694.

Tsay, H. S. and Tseng, M. T. (1979) Embryoid formation and plantlet regeneration from anther callus of sweet potato. *Bot. Bull. Acad. Sinica*, **20**, 117–122.

Tsay, H. S., Lai, P. C. and Chen, L. J. (1982) Organ differentiation from callus derived from anther, stem and tuber of sweet potato. *J. Agri. Res. China*, **31**, 191–198.

References

Villareal, R. L., Tsou, S. C., Lo, H. F. and Chiu, S. C. (1982) in *Sweet Potato. Proc. First Int. Symp.* (ed. R. C. Villareal and T. D. Griggs), AVRDC, Taiwan, pp. 313–320.

Withers, L. A. (1986) in *Plant Tissue Culture and its Agricultural Applications* (ed. L. A. Withers and P. G. Anerson), Butterworths, London, pp. 261–276.

Withers, L. A. and Williams, J. T. (1985) in *Biotechnology in International Agriculture Research*, International Rice Research Institute, Philippines, pp. 11–24.

Wu, Y. W. and Ma, C. P. (1979) Isolation, culture and callus formation of *Ipomoea batatas* protoplasts. *Acta Bot. Sinica*, **21**, 334–338.

Xin, S. Y. (1987) Studies on *in vitro* meristem culture and low temperature storage of sweet potato. *Crop Germplasm*, **2**, 34–36.

Xin, S. Y. (1988) *In vitro* storage of sweet potato germplasm. *Crop Germplasm*, **3**, 24–26.

Xin, S. Y. and Zhang, Z. Z. (1987) Explant tissue culture and plantlet regeneration of sweet potato. *Acta Bot. Sinica*, **29**, 114–116.

Xue, Q. H. (1987) Callus induction and plantlet regeneration of sweet potato cotyledon cultured *in vitro. Jiangsu J. Agri. Sci.*, **3**, 23–30.

Yamaguchi, T. (1978) Hormonal regulation of organ formation in cultured tissue derived from root tuber of sweet potato. *Bull. Univ. Osaka Pref.*, **B30**, 58–88.

Yamaguchi, T. and Nakajima, T. (1972) Effect of abscisic acid on adventitious bud formation from cultured tissue of sweet potato. *Jpn. J. Crop Sci.*, **41**, 531–532.

Yamaguchi, T. and Nakajima, T. (1974) in *Plant Growth Substances, 1973*, Hirokawa Publ. Co., Tokyo, pp. 1121–1121.

Yoshida, F., Kawaku, K. and Kakaku, K. (1970) Functions of calcium and magnesium on ion absorption by cultured free cells of tobacco and sweet potato. *Tamagawa Univ. Faculty Agri. Bull.*, **10**, 13–27.

8

Conservation of tree crops

C. P. WILKINS

8.1 INTRODUCTION

The human race is dependent on trees for timber, fuel, food, as raw materials for industry (fibre, pulp, oil, dyes, rubber), as raw materials for pharmaceuticals (enzymes, drugs), as beverages (tea, coffee, cocoa) and also for environmental stabilization. Forests provide a habitat for wildlife, whilst ornamental trees improve the quality of human life by providing an aesthetic input in our homes, gardens and streets, and also enhance the recreational facilities of our parks and countryside.

Many developing countries are economically dependent on exports of tree products to provide essential hard currency. For some countries, the economic situation is such that remaining hardwood timber reserves in primary forests are viewed as valuable assets that may aid in the repayment of interest on debts to foreign international banks.

In terms of tropical tree crops, coffee is the most important agricultural commodity on international markets, with a value of some 9–12 billion US dollars annually (Sondahl *et al.*, 1985). Coffee production is of necessity confined to the tropics, with seven major producing countries (Brazil, Colombia, Indonesia, Ivory Coast, Mexico, Guatemala and El Salvador) for whom annual coffee export revenues are a major source of foreign income. Other third-world countries reliant on tree products for a large percentage of their export earnings include Nigeria, which presently supplies half of the world's palm oil (Reynolds, 1982).

In addition to providing a valuable export commodity, tropical fruits (mainly plantation crops such as banana/plantain, papaya and date) provide an important component of the diet in some countries, and in some instances comprise the staple carbohydrate source. Also, aside from cash crops such as coffee, rubber, fruits, edible oil and timber, some

In Vitro *Methods for Conservation of Plant Genetic Resources. Edited by John H. Dodds. Published in 1991 by Chapman and Hall, London. ISBN 0 412 33870 X*

countries earn extra income from the extraction and sale of crude secondary products, such as the proteolytic enzyme papain, from papaya species, used in the food industry (Litz, 1985). Similarly, the tropical tree nut cashew (*Anacardium occidentale*), which ranks third after almonds and hazelnuts on the basis of total world production of tree nuts (IBPGR, 1986a), provides three valuable export products. These include the edible nut (the primary export item), an edible swollen pedicel that also provides a gum used in varnishes, and in addition a phenolic liquid extracted from the shell which has numerous industrial uses, including brake-linings for motor vehicles, paints, chemical and plastics.

Within developed nations, several major horticultural industries are tree-dependent. In temperate regions, production of Rosaceous tree fruits such as apple has been estimated at 35 million metric tons (FAO, 1980), whilst Krul and Mowbury (1984) have estimated world production of grapes at 37 million metric tons. Other examples of tree-dependent industries include the production of Christmas trees, for which the domestic market in the USA alone has been valued in excess of 375 million dollars annually (Durzan, 1985).

There is presently an ever-increasing demand for tree products and a concurrent growing awareness of the potential value of previously unexploited tree resources. However, in recent years, international concern over the uncontrolled destruction of natural forests has resulted in the global realization that such reserves are finite and are being progressively destroyed without consideration of the potential consequences involved.

Very recently, the subject of tropical deforestation has attracted intense media interest, and has become a topical 'green issue', being discussed amongst political parties and governments of many nations. Recent coverage in the popular press and by national television services of many countries have further served to highlight this issue, especially since the murder in December 1988 of Chico Mendes, the former leader of the rubber tappers in the Brazilian forest province of Acre, who campaigned against the continual destruction of the rainforest by farmers.

8.2 LIMITATIONS OF CONVENTIONAL TREE-PROPAGATION METHODS AND THE NEED FOR CONSERVATION

Figures relating to the rate at which tropical deforestation is progressing have become much quoted statistics. Lanly (1982) cited the (1980) figure of 11.3 million ha/year, whilst Cross (1988) provides a figure of 7.5 million ha/year, estimated as a result of a survey by the United Nations Food and Agriculture Organization. However, it was pointed out by the World Wide Fund for Nature (WWF) that the latter figure, which refers to closed tropical broadleaf forest, greatly underestimates the extent of the damage. For example, the survey did not take account of firewood gathering, excessive grazing, damage from fire and overlogging for both commercial and domestic purposes. The WWF suggests that the true figure is nearer 14.37 million ha/year. In any event, it is generally acknowledged that the world's total reserves of closed tropical broadleaf forest — the most important in the tropics — presently estimated at 1.2 billion hectares, are disappearing at a rate of between 0.6% and 1.5% each year.

Moreover, clearance of forests for logging, agricultural use, urban development and/or large development projects such as dams, roadways, mines and power stations, is predicted to increase in the near future (Lugo, 1985). Shortages of forest products have been forecast by the end of this century (Keays, 1974), and there is an urgent requirement for large numbers of improved, fast-growing trees of shortened rotation for pulp, paper, timber and furniture (Biondi and Thorpe, 1982). In addition, demands for wood and its derivatives will increase for use in the plastics industry and for fuel, either directly or indirectly (Karnosky, 1981). Already, in some arid regions of the world, there is a fuel wood crisis (IBPGR, 1983), because, although food is available, or has been provided through relief agencies, it cannot be cooked due to lack of wood (Eckholm, 1975; Durzan, 1985).

In countries with small areas of tropical forests, rates of deforestation can be very high, and are already at a critical stage (e.g. Nepal, Costa Rica, Ivory Coast, Haiti) (Lugo, 1985). Pearce (1989) has very recently drawn attention to the situation that exists in the Philippines, where trees are rapidly disappearing, and virtually all of the original pristine rainforest has now gone. Although some 20% of the country is technically forested, most of this land has been seriously damaged. The loss of forests in the Philippines is being repeated in nearby Sarawak in Malaysia. Furthermore, the logging companies, determined to meet the needs of Japan, the world's largest market for tropical timber, are poised to move into Burma, Laos, New Guinea (almost 80% of which is still forested) and the Amazon rainforest, which is currently being destroyed by farmers rather than loggers. The resultant deleterious effects of deforestation on the ecosystem are well documented: erosion of topsoil, plant denudation leading to species extinction, desertification and disturbances to the water table (Kong and Rao, 1982).

Rubeli (1989) cites the case of Thailand, which in November 1988 suffered its worst natural disaster for decades, when flooding and landslides killed more than 450 people. Recognizing that deforestation was the main cause, the Prime Minister, Chatichai Choonhaven, persuaded his cabinet to agree to terminate all existing logging concessions in the country. Recently, news media coverage throughout most developed nations has highlighted the international concern already expressed by conservationists and scientists, that deforestation is resulting in the extinction of valuable or potentially valuable forest tree species, as well as destroying unique wildlife habitats. The tropical timber industry destroys valuable and unexploited resources, including trees that provide fruits, nuts, gums and potential medicines. Pearce (1989) cites the results of a recent study concerning a small patch of Amazon rainforest, which found that fruits and latex represented 90% of the potential commercial value of the forest, and wood just 10%. Many workers have commented on the destruction of natural forests, and the need to explore, evaluate and conserve as much as possible of the remaining forest tree genetic resources for future generations (Sykes, 1975; Hawkes, 1976; Longman, 1976; Blake, 1984; Brown and Lugo, 1985; Cross, 1988; Pearce, 1989; Rubeli, 1989).

The causes of deforestation are multiple, ranging from logging and/or clearance for subsequent farm use (either cultivation or grazing) to stripping for development projects, often financed by International Aid Agencies. South and Central America possesses the largest area of rainforest, with 56% of the world total, the rest being split between south-east Asia and Africa (Cross, 1988).

Arguments persist between opposing groups of conservationists as to who or what is actually to blame for rainforest destruction. This topic has been discussed by Pearce (1989), who maintains that, although commercial logging is directly responsible for only a fifth of the loss, related activities such as road building, damage caused to trees as logs are hauled through the forest, and heavy machinery, all increase the toll. Perhaps more importantly, road-construction activities concurrent with logging operations serve to open up huge tracts of previously inaccessible forest to the landless poor of the tropics, who then move in to practise 'shifting cultivation'. However, these new colonists often lack the skills or determination to maintain the forest environment, and there is a consequent loss of fertility of the forest soil, leading to eventual destruction of the habitat.

It is universally agreed that internationally coordinated strategies for rainforest preservation must soon be devised and implemented. However, as Cross (1988) and Pearce (1989) both point out, there is a wide gulf between theory and reality concerning the struggle to save the world's dwindling tropical forests. There have been many, often controversial, suggestions concerning ways in which the loss of rainforests may be halted. Often, suggestions from different parties are contradictory, and in any case depend on the individual causes of deforestation within any particular country in question. As mentioned previously, logging is the primary cause of deforestation in south-east Asian countries, whilst farmers are mainly responsible for destruction of the Amazon rainforest. However, since Cross

(1988) has pointed out that exports of tropical timber from Brazil to Japan are predicted to rapidly increase up to the order of 40 000 m^3/month, the latter situation may soon change.

Some representatives of international agencies and research intitutes advocate stopping all Western-financed aid programmes for the Third World, and others propose setting aside enormous areas of virgin rainforest as protected reserves, with a total moratorium on logging. Yet another point of view specifies the need for sustainable management of forest resources, and maintains that a total ban on logging would be counter-productive, and do little to curtail forest destruction by farmers. Pearce (1989) quotes evidence from some Philippine islands on which logging is banned that shows higher rates of deforestation (3–5%/year) than in areas that are managed (where the average is 0.5–2%). However, the amount of forest presently undergoing sustainable management for producing timber is less than one million hectares, and is concentrated mainly in Trinidad, Tobago and Malaysia. Africa, South America and the main Asian timber-producing countries do not have a single proven example of sustainable management.

At present, it appears that no realistic solution to the problem of tropical deforestation yet exists.

In addition to concern over dwindling forest tree gene pools through deforestation, many scientists have also expressed grave concern regarding the urgent need to preserve threatened germplasm resources of other tree species, for example both tropical and temperate fruits (Zagaja, 1970, 1983; Lamb, 1974; IBPGR, 1986a). The widespread tendency to use only a few, high-selected introduced clones has resulted in a serious erosion of diversity amongst natural populations. The consequences of such a situation will be discussed in more detail later in this section.

Concurrent with the growing awareness that tree genetic resources must be conserved has been the realization that conventional methods of tree propagation, breeding and improvement currently in use have many serious limitations and shortcomings (Wilkins et al., 1985). It is obvious that one effective remedial measure to check or counter the damage caused by deforestation is the immediate implementation of reforestation pro-grammes (Kong and Rao, 1982). The objective is to replant cleared areas with native desirable tree species of commercial value so as to provide a readily 'cashable' commodity that will assure a financial return in 15–20 years.

Unfortunately, due to non-availability of seedlings when required, this has not proved possible. Lugo (1985) has proposed that much of the present demand for endangered primary forests could be reduced through the establishment of properly managed plantations or the intensive man-agement of secondary forest. Tropical tree plantations produce ten times more wood per unit land area and time than unmanaged natural forests, and produce about two to three times more wood volume than temperate tree plantations. It follows that taking action now to ensure that cleared areas of forest are effectively utilized, such as investment in the establishment of plantations, would pay dividends in the future, since the loss of natural wealth may be compensated for or replaced by additional revenue for the country (Raven, 1980). Such action could also (eventually) relieve some of the current pressure on natural forest reserves.

However, at present only one hectare of plantation is being planted for every ten hectares of forest that are cleared, and the area of tropical tree plantations is currently only about 0.6% of the area of tropical forests (Lanly, 1982).

In general, most forest tree species are seed-propagated, although some are asexually propagated as clones via rooted cuttings or grafting. Species such as *Populus nigra* (Lyr *et al.*, 1967), *Cryptomeria japonica* and *Chamaecyparis obtusa* (Thulin and Faulds, 1968) have been propagated by rooting of cuttings for several centuries. Propagation from rooted needle fascicles (also known as brachyblasts, short shoots or dwarf spurs) is a commonly used technique for multiplication of pine species (Girouard, 1971). However, in general, such methods of forest tree propagation are usually inefficient, and not normally applicable to most forest species, since older (i.e. physiologically mature) trees are extremely recalcitrant to these methods of propagation. In addition, repeated cloning over many years can lead to problems due to pathogen accumulation (Shields and Bockheim, 1981). Problems can also arise because clones are often highly site specific (Bonga, 1987), and clones of a selected superior tree may grow well only on sites similar to the original site of the selected parent tree. Several authors have given accounts of conventional methods of forest tree propagation, and the problems inherent therein (Mott, 1981; Biondi and Thorpe, 1982; Bonga, 1982a,1987; Brown and Sommer, 1982; Wilkins *et al.*, 1985).

With the seed-propagated forest species there are many obstacles to rapid genetic improvement. The optimal growth habit desired by foresters for such species is a prolonged juvenile phase, during which there is a rapid increase in height, since the seedlings must develop in the competitive shading of the parent canopy, or within the shrub layer of the forest floor. However, such a prolonged juvenile period before flowering and maturity is a handicap to forest tree breeding programmes, since trees must reach maturity before superior individuals can be identified with confidence (Cannell and Last, 1976). Present tree-improvement programmes therefore consist of an initial selection procedure to search for genotypes with desirable traits amongst natural stands, or first-generation hybrids. Once such 'elite' trees have been identified, the superior trees can be crossed in the hope of producing even more superior individuals among the seed progeny. However, since the elite trees may be several kilometres apart, and the flowers inaccessible, accomplishing such crosses is often extremely difficult and impractical (Mott, 1981). In addition, in the event that such crosses can be made, and the resultant seeds collected and planted, the seedlings take some 20–30 years to pass through the juvenile phase before they can be assessed and further crosses made.

A more usual and practical approach to tree improvement is firstly to select a mature parent tree which has been identified as producing a high proportion of desirable seed progeny. This tree is then used to establish a 'seed orchard', in order to obtain sufficient quantities of superior seed for use in reforestation. Scions are taken from the parent tree, and grafted to young rootstocks; a number of such composite clones are then planted to establish an orchard specifically for large-scale production of desirable seed. However, such orchards are both difficult and expensive to establish, requiring special care and attention for many years and taking some 10–30 years to become productive (Mott, 1981). This procedure also assumes that grafting methods are feasible for the species in question. In spite of these drawbacks, the economics of forest production are such that even gains as low as 2–5% are worthwhile, and justify the initial expenditure (Libby *et al.*, 1969; Carlisle and Teich, 1971).

Perhaps the most important group of tree species for which reliable methods of propagation are unavailable are the so-called 'luxury timbers'. A publication by the National Academy of Sciences (1979) has outlined some of the urgent research needs of these broadleaf species. Wilkins and Dodds (1983a) have recently discussed these crops with regard to conservation needs. One such threatened species is the tropical hardwood known as afrormosia (*Pericopsis elata*) or African teak. This species, native to various western African countries, provides one of the most valuable woods on world timber markets. As a consequence, most natural stands of this tree have been heavily cut. Since the natural regeneration of afrormosia is negligible and trees are not being planted on a large scale, the species is facing economic and biological extinction. Immediate action has been recommended in order to preserve afrormosia germplasm, with a proposal that Ghana, Cameroon and Zaire each set aside 2–3 km^2 of afrormosia forest as a conservation reserve, protected from exploitation but available to foresters for germplasm collection. However, this method of conservation has many inherent drawbacks, and, in addition, afrormosia seeds have a maximum storage life of only a few months. There is also an urgent need for research dealing with fundamental aspects of afrormosia silviculture. Apart from the fact that afrormosia seedlings require open sunlight and therefore will not grow beneath an already established canopy, foresters have virtually no knowledge of the flowering and seeding habits of the species, including: pollination methods, fruit ripening and frequency of seed years, seed germination, ecological requirements of young seedlings and the performance of trees under plantation conditions.

Similar proposals regarding conservation needs have also been made for various *Intsia* species, in an attempt to preserve remaining gene pools of this valuable genus. Already, in most countries few trees are left in natural stands, as the genus has been so extensively exploited throughout much of south-east Asia. Some countries (i.e. Malaysia) have already restricted the export of *Intsia* wood, but the complete decline of these species as economic plants is imminent. Again, as for afrormosia, little is known of these species regarding environmental tolerances, seeding and flowering habits and susceptibilities to pests and pathogens. Concern has also been expressed regarding various members of the two genera *Pterocarpus* and *Dalbergia*. Species of *Pterocarpus* include the timbers padauk, narra and muninga – some of the most valuable on international trade markets. None of these species is extensively cultivated, and native stands are fast disappearing; such species are also virtually untested as plantation crops. A similar situation is apparent in the case of the genus *Dalbergia*, some dozen or so species of which comprise the rosewoods. All accessible stands of these species are now gone and, apart from in Java and India, there are no plantations or trial plantings. Other tropical hardwoods for which reliable propagation methods need to be developed include teak (*Tectonia grandis*).

It will be obvious that an ideal method for propagating forest trees would be to clone superior elite trees by asexual propagation. Such elite trees could be either heterotic hybrids, or those specimens from field populations which have been evaluated for desirable characteristics such as height, straightness of trunk, wood and grain quality, growth rate, pulping characteristics, disease resistance and cold tolerance. By this means, superior genetic characteristics are propagated unaltered; that is, without the loss of exceptional gene combinations through gene recombination in the sexual cycle (Biondi and Thorpe, 1982).

For most trees, however, once maturity is attained vegetative propagation becomes either extremely difficult, impractical, or impossible. Propagation of trees from cuttings normally depends on the age of the parent tree, with rooting capacity of the cutting declining rapidly with increasing age of the parent plant. In addition, such factors as percentage rooting, speed of rooting, root length and number, survival, and growth rate all decline rapidly when the parent tree approaches maturity (Girouard, 1974).

Trees such as aspen (*Populus*), black walnut (*Juglans*), cherry (*Prunus*), sycamore (*Platanus*), beech (*Fagus*), eucalyptus (*Eucalyptus*), chestnut (*Castanea*), and most conifers are among those trees which characteristically root only from juvenile cuttings (Mott, 1981). Maturation is therefore one of the key obstacles to successful vegetative propagation of forest trees (Biondi and Thorpe, 1982). An additional complication often encountered when attempting to root cuttings of certain coniferous species is that the cuttings often vary drastically in their growth rates, and also sometimes assume an irregular, 'plagiotropic' growth form, in which rooted branch cuttings continue to grow with a horizontal orientation and bilateral symmetry (Biondi and Thorpe, 1982). Eventually, the terminal meristem changes to radial symmetry and vertical growth (orthotrophy). However, this eventual reversal to upright growth frequently displays intra- and interclonal variations and is thus erratic, thereby making the evaluation of genotypes in a selection experiment almost impossible (Libby, 1974). The occurrence of plagiotropic growth in newly rooted cuttings is usually dependent on the age of the parent tree and the original position of the cutting on the parent tree (Ross, 1975; Copes, 1976).

The often-cited classic example of plagiotropic growth is *Araucaria cunninghamii* (Hoop pine), where only cuttings from vertical stems will form normal plants after rooting. Rooted cuttings from primary and secondary branches, instead of growing vertically, have been observed to maintain a horizontal growth habit for over 50 years, with the rooted primary branch cuttings forming laterals, and with the rooted secondary branch cuttings failing to do so (Haines and De Fossard, 1977; Meins and Binns, 1979). Similar, but less extreme and persistent, trends have been observed in *Larix, Picea* and *Pseudotsuga* (Frohlich, 1961; Wareing, 1971), and also in some hardwoods (Frohlich, 1961). In douglas fir (*Pseudotsuga menziesii*), plagiotropism is already noticeable in rooted cuttings from seedlings only 5–7 months old (Ross, 1975).

Phenomena such as plagiotropism, when a new plant seems to retain the characters of the portion of the plant from which it came, have been termed 'expressions of topophysis' (Robbins, 1964). Comprehensive and interpretative reviews of juvenility in woody plants have been given by various workers (Doorenbos, 1965; Kozlowski, 1971; Brochert, 1976), and the subject has been discussed by Bonga (1982b) in relation to vegetative propagation.

A recent review by Durand-Cresswell *et al.* (1982) has discussed the limitations of presently available conventional methods of vegetative propagation when applied to members of the important dicotyledonous genus *Eucalyptus*. The review stresses the urgent need to develop reliable and efficient methods of vegetative propagation for this genus, species of which are in increasing demand for timber, pulp and paper production throughout the world. In plantations, this genus rates as one of the most productive forest crops; in addition, some species (i.e. *E. citriodora*) are a valuable source of essential oils. Mascarenhas *et al.* (1982) have reported identifying elite trees of *E. citriodora* whose leaves contain 3–5% oil.

Originally endemic to Australia, New Guinea and the Philippines (Cremer, 1969), the genus *Eucalyptus* consists of 445 species, 24 subspecies and 24 varieties (Chippendale, 1976). To date, species of *Eucalyptus* have been planted in South America, Africa, Asia, Spain, Portugal, Middle-Eastern countries and North America, and are adaptable to a wide variety of soil types, including sandy, saline, swampy, waterlogged, rocky and dry (Perron, 1981). In addition to the uses listed earlier, *Eucalyptus* trees are employed as windbreaks and firebreaks, and also for rapid covering of the ground after forest fires to prevent erosion.

However, when applied to *Eucalyptus*, most conventional methods of propagation have failed, especially when applied to adult tissues. As a consequence, various 'rejuvenation' techniques have been attempted in order to try to retard or reverse the maturation process and hence obtain rootable cuttings. One of the most widely employed methods of rejuvenation has been the use of grafting techniques, whereby adult (mature) tissues are grafted onto seedling (juvenile) rootstocks to rejuvenate the desired adult tissues and subsequently obtain rooted cuttings (Cauvin and Marien, 1978; Franclet, 1979,1981). However, although some ten different grafting and budding techniques have been tried for *Eucalyptus* (Hartney, 1980), this means of vegetative propagation is uneconomic, since delayed graft incompatibility (often after several years) may occur (Davidson, 1977).

An additional problem encountered during the propagation of *Eucalyptus* by grafting has been the occurrence of plagiotropic growth, as discussed earlier. For example, if lateral branches of *E. cladocalyx* are used as grafts the resultant growth is always horizontal (Maggs and Alexander, 1967).

Brown and Sommer (1982) have reviewed several additional techniques of rejuvenation that have previously been applied to various broadleaf forest species, in an attempt to obtain physiologically younger material (i.e. rejuvenated) for establishing clone banks to enhance the rooting response. These methods have included girdling of the main trunk just above the root collar, which frequently releases suppressed buds to form stump shoots that may revert to a more juvenile condition and be rooted with a reasonable degree of success (McAlpine, 1964). Similarly, partial stem girdles placed along the lower main trunk result in the release of suppressed buds to form epicormic branches which can frequently be rooted with success in mist beds (Kormanik and Brown, 1974). Garner and Hatcher (1962) and Libby (1974) have used the technique of severe pruning or 'hedging' to arrest the maturation phase for propagating woody material that is otherwise difficult to root. Hedging is commonly employed with coniferous species as an aid to rejuvenation.

A different situation exists regarding fruit-tree crops such as the Rosaceous temperate fruits. For these species, asexual clonal propagation is the norm, since they do not produce true-to-type seed and are therefore highly heterozygous. The methods of propagation currently in use for such crops are either budding or grafting, in which a single bud or short piece of shoot (the scion) is cut from the desired variety of tree that is to be increased, and is then applied to an already established tree (the rootstock) so as to form a composite tree. This indirect method of propagation is necessary because hardwood cuttings of most fruit trees are extremely difficult to root (Howard, 1978). Such techniques are therefore the only feasible way to obtain a number of clones of a desirable fruit-tree cultivar. An added complication of fruit-tree propagation is that most commonly used rootstock cultivars are also highly selected clones, the selection of which can have a profound effect on the precocity and productivity of the scion cultivar, conferring either dwarfing or vigour qualities on the scion and influencing such factors as tolerance to soil and climatic variables, resistance to soil pests and pathogens, yield efficiency, anchorage, and ease of propagation (Westwood, 1978). Rootstock cultivars are normally propagated by the techniques of stooling or layering. This process takes three years, and demands expensive nursery facilities and skills (Jones et al., 1979). Due to the slowness of these conventional methods of asexual propagation, many newly released cultivars from fruit-tree breeding programmes are often in very short supply. This applies both to rootstock and scion cultivars.

The practice of vegetative propagation of fruit trees as clones has been employed for many centuries. Written records of vegetative propagation date back more than 2000 years. Janick (1979) has discussed ancient texts by Theophrastus (written about 300 BC) and Pliny which discuss propagation of trees by various means, including cuttings, grafts, suckers and layering. The ancient Greeks and Romans described several varieties of apples, plums, grapes and cherries, and were conversant with methods of propagation by both budding and grafting (Sekowski, 1956; Zielinski, 1955). Similarly, peaches and apricots were initially domesticated in China, where their cultivation was known from at least 2000 BC (Sekowski, 1956). In some instances, clones established in antiquity have survived until the present day. The grape cultivar Cabernet Sauvignon has reputedly existed since Roman times (Mullins and Srinivasan, 1976).

Although conferring enormous benefits on fruit growers, the use of sophisticated techniques of fruit-tree clonal propagation also possesses several inherent disadvantages. As discussed earlier with regard to certain forest species, clonal propagation of fruit trees, especially on a large scale, or involving the maintenance of specific clones over long periods of time, carries certain risks. Clonal populations have little adaptability to site variables, and are often severely affected by pests and diseases.

One of the most serious consequences arising from the use of clonal methods of propagation has been the widespread dissemination of fruit-tree viruses. As discussed by Hudson (1982), such viruses can have many effects on the size, colour and quality of fruits and as such are fairly easy to recognize. However, other viruses are more insidious in their effects, and, whilst not inducing obvious visual symptoms, may affect the growth, vigour, productivity and longevity of trees. In some instances, virus accumulation has resulted in clone deterioration. Passecker (1964) has discussed the example of the apple clone Roter Stettiner, which for a 500–600-year period was one of the most productive clones in central Europe, but has now become almost extinct because of gradually increasing degeneration problems. More recently, Westwood (1978) has described how certain cultivars of morello cherry (*Prunus cerasus*) can act as symptomless carriers for necrotic rusty mottle virus. Similarly, many apple cultivars show no symptoms when infected with apple stem grooving virus (SGV), although the sensitivity of other commercial cultivars has resulted in international quarantines against the sale of plants that cannot be certified as free from infection (Huang and Millikan, 1980). This situation is especially serious for fruit-tree growers, since the establishment of a fruit-tree orchard involves a high capital expenditure and is expected to give financial returns over some 20 or more years of production.

As a result of concern over the spread of virus-infected material, the EMLA (East Malling: Long Ashton) scheme was developed for fruit-tree stocks and scions. Material designated as being of EMLA quality is certified as virus-indexed and therefore allows the production of first-class virus-indexed fruit trees. However, the problem remains that all such highly selected stocks are asexually propagated as clones, and, as for example in the case of the dwarfing apple rootstock M9, are slow and difficult to propagate by conventional means, and are therefore in constant short supply. A similar situation exists regarding *Prunus* rootstocks, such as the plum rootstock Pixy (*P. insititia*).

The production and widespread availability of highly selected and productive fruit-tree cultivars has resulted in the utilization of fewer main cultivars of any particular fruit crop, especially since the use of different rootstocks serves to extend the usefulness of any particular scion cultivar. For example, although some 8000 apple cultivars have been described, in the USA only two cultivars (Delicious and Red Delicious) and their mutations account for over 50% of total crop production. A comparable situation exists in other temperate fruit-producing countries such as Italy, West Germany, France, Japan, China, The Netherlands, Canada and Great Britain. Unfortunately, this widespread tendency to use only a few, highly selected, uniform, high-yielding cultivars has resulted in a serious decline in the available genetic diversity amongst fruit-tree populations.

In addition, over the last two decades or so, the widespread introduction of modern high-yielding fruit cultivars, together with modernization of fruit-production practices (i.e. mechanization, intensive regimes of pesticide/herbicide application and use of specialized rootstocks), in those countries that were formerly reservoirs of fruit-tree genetic diversity, has resulted in serious erosion of indigenous fruit-tree gene pools. The main centres of genetic diversity of the temperate-zone tree fruits were first described by authors such as Vavilov (1926, 1930) and Zielinski (1955). More recently, Zagaja (1970) discussed the vast range of temperate tree fruits to be found in just one area of genetic diversity, namely Turkey. Early surveys had revealed a vast range of available genetic variation, including adaptations to specific climatic conditions such as extreme frost resistance in certain apple species, pathogen-free populations of sweet cherries, variation in flowering period among almonds, dwarf varieties of *Prunus* species and adaptations to highly calcareous soils. However, the author emphasized the point that all such tree fruits in Turkey were threatened with imminent extinction, due mainly to modernization of fruit production within the country. For example, apricots, which were previously propagated by seed, were being exclusively propagated by grafting. Introduced peach cultivars had become so successful as to have almost entirely eliminated local varieties, and recent introductions of standard apple, pear and cherry cultivars were having a similar effect.

The seriousness of this situation is being exacerbated in some countries, since although in some cases (i.e. as in Turkey) such introduced species/cultivars prove to be outstanding, in other cases the introduced varieties rapidly succumb to adverse climatic variables, or pests and pathogens. Most highly bred cultivars perform exceedingly well, but only under regimes of cultivation involving sophisticated agronomic technique. For example, the apple cultivar Golden Delicious (an important apple variety of most major fruit-growing countries) has a high crop value, but the normal programme of disease and pest control requires the application of up to 20 different sprays per year (Hudson, 1982). It follows, therefore, that in order to introduce and utilize such a variety in a developing country, one must also import the appropriate agronomic expertise. In many developing countries this is impossible, since the structure of the agricultural economy effectively precludes the grower (usually a poor farmer) from obtaining the necessary capital or credit to implement such technology. Several workers have thus stressed that it would be far better not to import species that rely on sophisticated agronomic techniques to achieve good production, but instead to extensively screen and evaluate existing germplasm collections to find species/cultivars more suited to the particular ecological and pedological conditions, and resistant to the indigenous pests and diseases: in other words, to exploit the existing genetic potential of a crop.

Various workers have pointed out that, in some cases, entire selections of wild fruit-tree germplasm have already been irretrievably lost. These might have been of use in future breeding programmes and might have possessed genes for useful characters such as outstanding disease or insect resistance, nutrient uptake efficiency or hardiness. Brown (1975) has discussed the use of wild apple species in a breeding programme designed to introduce resistance to apple scab (*Venturia inaequalis* Wint.) into commercial apple cultivars. More recently, Fogle and Winters (1980) have cited an example in which large collections of pear germplasm were discarded after failure of accessions to show resistance to bacterial fire blight (*Erwinia amylovora*). However, the collections were not screened for resistance to other diseases such as 'pear decline', a new disease which has already eliminated susceptible *Pyrus ussuriensis* and *P. pyrifolia* stocks, and is caused by an insect-borne mycoplasma. In addition, the potential value of wild related species as future rootstocks with special features such as dwarfing or tolerance to adverse soil conditions (Westwood, 1978) or as pollinators for commercial cultivars (Crassweller *et al.*, 1980) is only just being realized.

There is evidently a pressing need for conservation of fruit-tree germplasm resources on a worldwide scale. As well as conserving ancient genotypes, locally selected cultivars and wild species of fruit trees, there is also a need to conserve current and recently obsolete cultivars of fruit-tree scion varieties, and highly selected rootstock clones. Since some of these rootstock clones are sterile, e.g. the semi-dwarfing hybrid cherry rootstock Colt (*Prunus avium* × *P. pseudocerasus*), storage in a vegetative form is a necessity. Even where possible, seed storage of such species is obviously unacceptable, since varietal identity will not be maintained due to the inherent heterozygosity.

A recent report by the International Board for Plant Genetic Resources in Rome (IBPGR, 1985) has demonstrated that storage of seed (when produced) of temperate fruit species is technically feasable using conventional seed-storage techniques (i.e. low temperature and reduced moisture content). Seed storage is theoretically the most effective means of conserving the genetic diversity of a species. However, the long juvenile phase typical of temperate tree species hinders the replenishment of seed-bank stocks when fresh supplies of viable seed are required. This procedure necessitates growing trees for seed production, an expensive process in terms of both space and labour requirements.

It is obviously far better in the case of clonally propagated crop species to conserve material vegetatively, so as to maintain the genetic integrity of individual fruit-tree genotypes. However, there are numerous problems associated with the conservation of fruit trees by conventional means, usually in the form of orchards (also referred to as 'field gene banks', 'clonal repositories' or 'living collections'). These include space and labour considerations, and the need to sample effectively and conserve as much as possible of the genetic variation contained within a population. This aspect will be discussed in more detail later in this section. Such collections are also prone to possible catastrophic losses due to: (1) attack by pests and pathogens, (2) climatic disorders, (3) natural disasters (e.g. earthquakes), and (4) political/economic causes.

In spite of the high cost of maintaining such collections, many National and International fruit-tree germplasm repositories have been established (Brookes and Barton, 1977) and the value of such collections has been amply stressed (Thompson, 1981). Furthermore, a comprehensive inventory of available North American and European fruit and tree nut germplasm resources has recently been compiled (Fogle and Winters, 1980). The latter inventory is a computer listing of fruit clones held worldwide in the collections of institutional and private breeders and variety testers, amateur pomologists, innovative growers and nurserymen. Some of the collections are, however, transient, since they are maintained as an integral part of breeding programmes, and are in danger of being discarded once the programme is terminated. More recently, Bettancourt and Konopka (1989) have compiled a directory of germplasm collections currently maintained by Agricultural Stations and Research Institutes throughout the world, and covering some 40 genera of temperate fruits and tree nuts.

Although world production of tropical fruits exceeds that of temperate fruits (FAO, 1982), less progress has been made to date concerning the development of reliable methods of vegetative propagation. As with temperate fruits, cultivation of the major important tropical fruit crops has been practised for several thousand years. The mango (*Mangifera indica*), for example, has been prized in India for some 4000 years (De Candolle, 1889). This country is also the major producer, accounting for 62% of the world crop. Both ancient and modern cultivars of the perennial tropical tree fruits have been derived from seedling trees that originally resulted from uncontrolled pollinations (Litz, 1985). Apart from species such as *Citrus* spp., and *Eugenia* spp., which reproduce asexually through the formation of adventitious embryos, methods of vegetative propagation have been employed for thousands of years to preserve the unique horticultural characteristics of highly prized tree selections that would otherwise be lost through gene recombination in the sexual cycle.

Despite the importance of tropical fruits as export commodities of many non-industrialized countries, there have been very few genetic studies that involve these plants. The genetic complexity of these trees, together with a juvenile period that lasts some 6–15 years, has meant that conventional plant-breeding approaches have had relatively little impact on cultivar improvement. In addition, under tropical conditions of high temperature and humidity, pathogens, pests and environmental stresses are constant

165

factors that curtail crop production. Also, the intensive commercial mono-culture of a few cultivars of tropical fruit trees over large areas has sometimes proved disastrous, due to overall crop susceptibility to diseases often caused by a single pathogenic organism. Litz (1985) cites the out-break of papaya ringspot virus in the Caribbean region as an example of such a situation.

Many authors have cited specific examples of tropical crops for which vegetative propagation poses many problems. With rubber (*Hevea brasiliensis*), for example, the present practice of vegetative propagation involves the planting of unselected seedlings for the root system, then budding the desirable trunk clone and, where applicable, budding again for the desirable canopy. This procedure has the disadvantage that budding is a skilled labour-intensive operation, and such labour will soon be in short supply in the principal rubber-producing countries such as Malaysia (Wan Abdul Rahaman *et al.*, 1982). Conventional techniques of plant breeding and improvement, e.g. selection from controlled pollinations, have also proved unsatisfactory when applied to rubber, due to extremely low (1–5%) fruit set.

The possible benefits of developing reliable methods for clonal propaga-tion of mature fruit trees could have a tremendous impact on tropical fruit production. Mascarenhas *et al.* (1982) have reported that elite trees of tamarind (*Tamarindus indica*) have a fruit yield five times higher than average, unselected trees (200 kg per tree as against 30–40 kg). However, Wilkins and Dodds (1983a) have also pointed out that, although the potential of the tamarind is just being realized, the species (as with many other tropical fruits) is in urgent need of genetic conservation. Germplasm collections need to be made in a region that stretches from Senegal across sub-Sahelian Africa to the Sudan, and also in India, Thailand and south-east Asia. Many scientists have stressed the urgent need to collect and conserve dwindling germplasm resources of other tropical tree crops such as cocoa (*Theobroma cocao*). Cocoa scientists have recently (Williams, 1982) agreed that collections of *criollo* types must be made throughout Central America, and also throughout the entire Amazon basin for amazonia types. It was noted that genetic variability observed during expeditions in 1973 has subsequently disappeared in less than a decade.

With other crops, such as the mangosteen (*Garcinia mangostana*), which many people consider to be one of the best-tasting fruits in the world, the lack of a reliable method of asexual propagation, together with very poor seed viability (Hanson, 1984), has prevented commercial exploitation. For these reasons, the mangosteen is virtually unavailable in what could be its major markets – Central America, South America, Australia and Africa (National Academy of Sciences, 1975).

Even in those tropical fruit species for which methods of vegetative propagation are available, such techniques are often slow and laborious. The pomegranate (*Punica granatum*), for example, may be propagated by means of air-layers, although trees employed for this purpose must be at least five years old, and a maximum of 70–100 layers may be obtained per tree (Mascarenhas *et al.*, 1982). Other tropical fruit crops for which vegetative propagation poses problems include papaya (*Carica papaya*). Plants of *Carica* spp. may be either dioecious, or highly inbred hermaphroditic cultivars such as the Solo types of Hawaii (Litz, 1985). All papayas are propagated by seed (although rooting of cuttings is feasible). However, commercial plantings of dioecious papaya types usually contain some 30% of non-productive male trees, and there has therefore been a need to develop a method for propagation of elite female or bisexual trees.

A recent report commissioned by the International Board for Plant Genetic Resources (IBPGR, 1986a) dealing with genetic resources of tropical and subtropical fruits and nuts has cited many examples where erosion of valuable germplasm of these species and their wild relatives is already known to have occurred.

In the case of mango (*Mangifera indica*), genetic erosion is already serious in India, where diversity is greatest, with over 1000 unique cultivars. The aforementioned report points out that the mango is an important crop which has been assigned high priority by the IBPGR, and the species possesses much potential for the development of future trade and exports. It is also of importance because it grows on marginal soils (Sastrapradja, 1975). Although many old cultivars of mango exist in countries such as Pakistan, Bangladesh and India, little use has been made to date of available mango germplasm in improvement programmes. Mukherjee (1985) recently described 40 hitherto little-known species of *Mangifera*, with some remarkable genetic characteristics that could prove of great value if included in future breeding programmes.

With the mangosteen (*Garcinia mangostana*), Sastrapradja (1975) considered that there has already been severe genetic erosion of available mangosteen germplasm reserves in south-east Asia. Trees of mangosteen occur naturally in south-east Asian rainforests, and are hence under considerable threat, as this habitat continues to be destroyed. The report by Sastrapradja (1975) also concluded that genetic erosion in the cultivated durian (*Durio zibethinus*) was extensive in Indonesia, Malaysia and Thailand, and had also occurred to a lesser extent in Vietnam and the Philippines. There has also been a loss of potentially valuable related wild species in Indonesia and Malaysia. The major cause for this loss of germplasm was attributed to the widespread distribution of relatively few clonally propagated cultivars to farmers. In addition, Choke (1973) discussed the Malayan system of managed forest regeneration with regard to declining resources of *Durio* spp. This practice involves the selective removal of native fruit trees, including *Durio* spp., in regenerating forest stands to allow the unhindered growth of trees of high timber quality (and associated high commercial value).

Serious genetic erosion of fruit-tree germplasm resources, due to a tendency of farmers to utilize only a limited number of highly selected clones, has also been documented for the Japanese persimmon (*Diospyros kaki*). The number of cultivars grown in Japan has declined from about 50 to 10–15 over the last few decades, and, in addition, many of the 1000 or so previously known non-commercial cultivars have now been lost (Kajiura, 1980). A similar situation has been observed in the case of both langsat (*Lansium domesticum*) and guava (*Psidium guajava*), where traditional cultivars have been replaced by a limited number of selected clones. Sastrapradja (1975) has discussed the extensive genetic erosion of cultivated langsat throughout Indonesia, Malaysia, Vietnam and Thailand.

In contrast, continual deforestation of natural forest stands in South America, where wild relatives of avocado (*Persea americana*) are indigenous, has been responsible for loss of valuable germplasm. The species *P. theobromifolia*, which is potentially important as a blight-resistant rootstock for the cultivated avocado (IUCN, 1978), is already classified as 'endangered'.

In the case of the pistachio nut (*Pistacia vera*), genetic erosion of wild stands in western Africa and the Mediterranean countries has been attributed to a number of factors, including land clearance, charcoal burning and overgrazing by goats. Several wild *Pistacia* species are of potential value for use in future improvement work as rootstocks or pollinators.

A series of recent reports (Anonymous, 1982,1983,1984,1986) has described an extensive list of forest fruit species that occur in East Africa, south-east Asia and Latin America, that are sorely underexploited at present but which could prove of value in the future.

In the case of the arborescent monocotyledonous palm species, members of which comprise some of the most economically important tropical plantation crops, many problems exist regarding presently available conventional propagation techniques. Species such as date palm (*Phoenix dactylifera*) have been under human cultivation for some 6000 years (Zohary and Spiegel-Roy, 1975). For both the date palm and sago palm (*Metroxylon* spp.), vegetative propagation is possible from offshoots or suckers arising from axillary buds at the base of the stem. However, the number of suckers produced is very small, and, consequently, the achievable rates of multiplication are low (Choo *et al.*, 1982; Reynolds, 1982). Only one or two offshoots are usually produced during the entire life of any individual palm. Additional problems associated with this practice of propagation include the possible distribution of insect pests and diseases, as well as transportation difficulties, since offshoots may weigh in excess of

16 kg each (Reynolds, 1982). Seed propagation of date palm is possible but impractical, for various reasons. *Phoenix dactylifera* is dioecious and completely heterozygous in both male and female parents. There is much segregation in seed-derived populations, and subsequently varietal identity cannot be maintained. Seedling populations also possess a high percentage of non-productive staminate trees which cannot be identified until flowering. In addition, seedlings do not start to flower until 12–15 years of age, whilst clonal offshoots are capable of flowering within 5–6 years. Optimum fruit set in date palm requires artificial fertilization of pistillate trees. However, a metaxenic effect has been identified in date palms, necessitating selection of specific trees for use as pollen sources, in addition to selected seed parents (Reynolds, 1982).

For other palm species such as oil palm (*Elaeis guineensis*) and coconut palm (*Cocos nucifera*), vegetative propagation using conventional techniques is impossible. Coconuts are grown in virtually all tropical areas of the world as a subsistence crop. However, since seed-produced plants are genetically very variable, production levels are often very low and several countries cannot produce sufficient nuts to meet the needs of the home market (Blake and Eeuwens, 1982). As it has been estimated that elite coconut palms can produce 400 nuts per year compared with the average figure of 35, whilst selected oil palms produce some 30% higher yields than average, the potential benefits of developing reliable systems of clonal propagation are obvious.

With the oil palm, there has been a need for clonal propagation of female-infertile *pisifera* types, which are highly desirable sources of pollen for the production of high-quality oil palms of the *tenera* variety (the oil palm of commerce). Since such palms obviously do not produce seed, asexual propagation is the only feasible method of multiplication (Nwankwo and Krikorian, 1983).

Several workers have stressed the urgent need for germplasm conservation of the indigenous genetic resources of certain palm species in some countries. For example, Tisserat *et al.* (1981) have stated that native date palm (*Phoenix dactylifera*) genomes of Algeria and Morocco are bordering on extinction due to Bayoud disease (*Fusarium oxysporium* Schlect. var. *albedinis*). However, conservation of clonal material by conventional means would be virtually impossible, since there is no efficient method of vegetative propagation. Similarly (but for different reasons), scientists such as Iyer (1982) have stated that, in the case of coconut palm (*Cocos nucifera*), massive replanting programmes are threatening to destroy native genetic resources in the Philippines, Indonesia, India and Malaysia. The report points out that some scientists (Carlos, 1979) have expressed the need to proceed with caution, indicating that certain business-motivated agencies are attempting large-scale replanting in many countries with the same hybrids. Since vegetative propagation of coconut palm is not feasible, and storage of nuts (even if possible) would pose obvious difficulties due to space considerations, the report stresses the need to develop reliable methods of storage to attempt to conserve remaining germplasm resources, and suggests the use of embryo-culture techniques as a possible alternative. Williams (1982) has also commented on the urgent need for programmes to be initiated for the intensive collection of coconut germplasm throughout south-east Asia and the Pacific, due to the rapid loss of genetic variability within these regions, and the need to incorporate a wider genetic base into breeding programmes. The author also points out that little attention has so far been given to this urgent task, due to the cost and practical problems involved in sampling and transporting large quantities of nuts.

Due to reasons briefly outlined earlier in this chapter, the case for conserving threatened genetic resources of trees has been stated many times. However, there are many factors that hinder the effective implementation of programmes of collection, evaluation and conservation of tree genetic resources. Most important tree crop species fall into two categories (Simmonds, 1982): (1) woody, perennial outbreeders that are not, or cannot yet be, cloned, with characteristically short-lived seeds (termed 'recalcitrant') which cannot yet be stored; and (2) clonally propagated woody outbreeders, which are highly heterozygous, and for which seed propagation is either undesirable or impossible. Several reports by King and Roberts (1979b,1980) and Roberts and King (1981) have reviewed those woody species whose seeds are termed 'recalcitrant' and which cannot be stored by conventional techniques. Although some progress has been made recently regarding the storage of seeds of some so-called 'recalcitrant' species, for example *Citrus* (King and Roberts, 1980) and oil palm (Grout *et al.*, 1983), it appears that the development of reliable methods of seed storage for the majority of such species will not be attainable (Hanson, 1984). In addition, even for those tree species for which seed storage is technically possible, for example the temperate fruits (IBPGR, 1985), Hawkes (1982) has pointed out that the process of replenishment of seed stocks which becomes necessary when viability of stored seed falls below an acceptable level would require actual growing of trees (to maturity), with a consequent regeneration cycle of several years (5–8 years in the case of apples and pears). Moreover, with some tropical fruit or timber species, the time taken for seed regeneration would range from 15 to 80 years (Hawkes, 1982), and hence the longer the regeneration cycle, the greater is the proportion of the seed bank in the form of living plants rather than seed. It thus becomes obvious that with the tropical timber trees, for example, some of which take 50–80 years to attain maturity and seed-bearing status, any seed bank would effectively become a plantation, thereby obviating the need for seed storage, since the regeneration cycle would be longer than the storage cycle.

Obviously, all clonally propagated woody species must be stored in a vegetative form, and Hawkes (1982) has discussed the use of conventional methods for the conservation of such species. These include the establishment of *in situ* biosphere reserves, or nature reserves, 'gene parks', etc., which would preserve areas of natural vegetation in their original habitat together with the species which they contain. However, such biosphere reserves have limited application, since they could not hope to conserve the amount of genetic diversity which it would be desirable to preserve. The genetic diversity contained within a wild species can be partitioned into three main categories: (1) geographical or climal diversity; (2) interpopulation diversity in each particular sampling area; and (3) intrapopulation diversity, or the diversity to be found in each local gene pool.

Since such *in situ* biosphere reserves must of necessity be small in comparison to the total distribution of a species, such reserves would only be of use in conserving intra- and interpopulation diversity. For a widespread species, as most woody species are, they could not possibly be numerous enough to conserve geographical variation effectively. Further, even if *in situ* reserves could be established, they would be constantly under threat from epidemics, changing government policies, timber needs and the rapid spread of urban developments. In spite of these drawbacks, natural stands of wild trees have been maintained in several countries, including the USSR (Zagaja, 1983).

An alternative solution to the problems of conserving long-lived perennial woody species is the establishment of plantations and fruit-tree orchards. These would probably prove most effective in the tropics, where land space is readily available and labour costs are low. In developed countries, orchards or 'field gene banks' would require constant management, and would hence be costly to maintain. In addition, such collections would constantly be vulnerable to earthquakes, floods, drought, frosts, pests and diseases, and would also be at risk of the changing whims of government policies/parties, as well as the pressures of urban developments and wars. Another serious drawback to the use of orchard/plantation collections as mentioned earlier with regard to temperate fruits is the large number of individuals required to provide an adequate sample of genetic diversity, and the consequent large land area required for such a collection. Burley and Namkoong (1980) have estimated that the number of individuals needed to provide an adequate sample of the genetic diversity of a tree species would vary from 20 to 30 for a small single population, up to several hundred for gene-pool conservation and upwards of 5000 for maintenance of heterozygosity. Several authors have commented on the areas of land required to maintain collections of various tree species. Sykes (1975) has pointed out that even sampling one tree in every 1000 of the three million almond trees in Turkey would provide 3000 single-tree accessions, and would require an area of some 15 ha. Lamb (1974) estimated that 16 ha would be required for a collection of apple germplasm which includes cultivars, rootstocks, species and ecotypes. Similarly, the total area required for a germplasm collection of wild species, ecotypes and cultivars of pistachio has been calculated by Maggs (1973) at 21 ha.

As part of a global network of genetic resource activities, the IBPGR has recently compiled a comprehensive directory of germplasm collections of tropical and subtropical fruits and tree nuts held worldwide by scientific research institutes and germplasm resource centres (Gulick and van Sloten, 1984).

It is botanically incorrect to classify bananas and plantains (both *Musa* spp.) as trees, since the stem or 'trunk' of these monocotyledonous Angiosperms is a non-woody pseudostem composed of leaf sheaths. However, these crops will be considered in this chapter because members of the genus include plantation crops of major economic importance, and are a main dietary component (as carbohydrate) in some tropical areas. In addition, the conservation needs of these species are not considered elsewhere in this volume. Bananas (*Musa* group AAA) are normally utilized for dessert consumption, whilst plantains (*Musa* groups AAB and ABB) are usually cultivated as a starchy staple crop. Some workers have described members of the genus *Musa* as 'herbaceous trees' (Litz, 1985) or 'non-woody arborescent monocots'.

All commercially cultivated *Musa* types are sterile triploids, and hence must of necessity be propagated vegetatively, usually from suckers. However, such methods are laborious and time-consuming as far as producing significant quantities of homogeneous plants is concerned (Banerjee and de Langhe, 1985). Also, conventional methods of propagation are slow. All modern *Musa* cultivars have arisen over several hundred years of cultivation through somatic mutation within existing clones (Litz, 1985). There is therefore a high degree of genetic vulnerability to disease and environmental stress (Rowe, 1984). Compelling reasons for the need to develop reliable techniques for conservation of *Musa* germplasm have been given by de Langhe (1984).

Over the last few years, interest has been shown in multiple-use woody species. These species would be planted in poor-quality sites requiring little care and management, and hence would assist in land reclamation, being grown on steep hillsides, in low-nutrient and toxic soils, or in arid zones and tropical highlands. The selection criteria for trees in this category include wide adaptability, nitrogen-fixing ability, rapid growth (short rotation), ability to coppice, drought tolerance and the production of wood that burns without sparks or smoke. Species within this group include members of the genera *Acacia, Albizia, Prosopis,* and *Leucaena* (Durzan, 1982). A report of the National Academy of Sciences (1979) has indicated that crops like *Leucaena* can add up to 500 kg of nitrogen to the soil per hectare per year. A later report by the same agency (National Academy of Sciences, 1984) has discussed the value of *Casuarina* species for colonization of poor sites, dune stabilization, establishment of shelterbelts and production of fuelwood. Sattaur (1983) has discussed the replanting of large areas of Indonesia with the lamtoro tree (*Leucaena leucocephala*). The report states that some 42 million hectares of land in Indonesia are in need of regreening, and large areas of land are in a critical condition due to soil erosion. In plantations in the Philippines, the lamtoro tree provides 30–40 m^3 of biomass per year, and is one of the fastest-growing trees in the world. In addition, the tree is capable of growing in many different soils, from heavy clays to coral. As well as providing environmental stabilization, and assisting the establishment of other plants, the lamtoro provides a fuel with a calorific value equal to that of coal, and also makes excellent charcoal. A review by Palmberg (1984) has discussed the available genetic resources of tropical species that are of value as fast-growing fuelwood species, and also points out that some 1500 million people in developing countries are dependent on fuelwood for daily food preparation. Most such species are fast-growing leguminous trees possessing 'weedy' qualities.

8.3 TISSUE CULTURE OF TREE CROPS

8.3.1 Historical development

Since Gautheret, over fifty years ago, first demonstrated the feasibility of culturing woody tissues (Gautheret, 1934,1937), there has been enormous interest in the possible practical applications of tissue-culture techniques to solve some of the problems affecting programmes of tree propagation and improvement. A later report by Gautheret (1940) described the first successful induction of *in vitro* organogenesis (i.e. bud formation) in a woody species (*Ulmus campestris*). The early pioneering work of Gautheret was continued and expanded in several reports by Jacquoit (1949,1951,1955a,b), who obtained adventitious bud formation in callus cultures of both elm (*Ulmus campestris*) and birch (*Betula verrucosa*). At around the same time, Ball (1950) reported inducing the first buds *in vitro* from a callus culture of a Gymnosperm (i.e. *Sequoia sempervirens*).

Shortly after the latter reports by Jacquoit appeared in the literature (in 1955), it was recognized that the development and application of reliable tissue-culture techniques could aid enormously in forest tree improvement programmes, and several early workers (Geissbühler and Skoog, 1957; Haissig, 1965) stressed the potential of such an application.

Mathes (1964) was probably the first researcher to achieve differentiation and growth of both shoots and roots in callus cultures of a woody plant (*Populus tremuloides*); although, as Brown and Sommer (1982) have pointed out, no indication was given that the shoots and roots were connected to form plantlets. Later work by Wolter (1968) showed that shoots could be initiated on callus of *Populus tremuloides* using only a cytokinin (benzylaminopurine), and also that these shoots could subsequently be rooted on medium devoid of growth regulators. Shortly after, Winton (1970,1971) demonstrated that normal, free-living tree plantlets could be obtained from *in vitro* callus cultures of aspen. It was not until some five years later that Sommer *et al.* (1975) reported obtaining the first Gymnosperm plantlets from *in vitro* cultured excised cotyledons and hypocotyls of *Pinus palustris* (longleaf pine).

Concurrent with the development of tissue-culture techniques for achieving plant regeneration/propagation from *in vitro* cultures of forest trees, various reports began to appear in the literature describing the *in vitro* culture of other woody species, most of which related to commercially important horticultural crops, such as the temperate woody fruit species. Some of the early reports involving tissue culture of fruit crops were concerned primarily with the use of meristem-culture techniques to eliminate systemic virus infections from species such as gooseberry (Jones and Vine, 1968) and apple (Walkey, 1972). Other workers centred their investigations on the effects of cytokinin compounds on apple shoot tips (Jones, 1967; Pieriazek, 1968). Several later workers showed that it was possible to culture shoot tips of apple *in vitro*, but there was no multiplication of shoots and few rooted (Dutcher and Powell, 1972; Elliot, 1972; Quoirin, 1974). A later report (Abbott and Whitely, 1976) showed that shoots of apple seedlings and of the scion cultivar Cox's Orange Pippin could be induced to multiply *in vitro* by about ten-fold per month. However, shoots were small, and difficult to root, and plants were not established *in vitro*. Shortly after, Jones *et al.* (1977) demonstrated that virus-free shoots of the apple rootstock M26 could be rapidly multiplied and rooted *in vitro*, then subsequently established under non-sterile conditions with a high degree of success. Such work indicated that more than 60 000 plantlets could be produced from a single shoot tip over an eight-month culture period. Subsequent work by Jones and Hopgood (1979) applied similar techniques to the plum rootstock Pixy (*Prunus insititia*) and the cherry rootstock Mazzard F12/1 (*P. avium*). Both of these rootstocks were in short supply, since both were difficult to propagate by conventional means (Howard, 1978; Feucht and Dausend, 1976). In the same year,

James and Thurbon (1979) described the *in vitro* propagation of the dwarfing apple rootstock M9. This rootstock is widely used because of its effects on precocity and the control of tree and fruit size (Rogers and Beakbane, 1957; Tubbs, 1967). However, as with the *Prunus* rootstocks, cuttings are extremely difficult to root using conventional methods; softwood and hardwood cuttings give rooting percentages of 36% and 6% respectively (Howard, 1978). Numerous workers later applied micropropagation techniques to a wide range of temperate fruit genera (Wilkins and Dodds, 1983b; Zimmerman, 1985).

In contrast to the rapid advances made regarding the application of tissue-culture techniques to temperate fruit species, progress in the development of methods of *in vitro* propagation for tropical fruit trees has been comparatively slow. Early workers were quick to report the successful induction of organogenesis in callus cultures of *Citrus* (Chaturvedi and Mitra, 1974) and the early use of *in vitro* grafting techniques to obtain virus-free plants (Navarro *et al.*, 1975). Initial progress was also made in obtaining plantlets from tissue cultures of papaya (Yie and Liaw, 1977). With most other important tropical fruit species, however, progress has been extremely slow and it is only during the last five years that many such species have been successfully introduced into *in vitro* culture, e.g. mango (Litz *et al.*, 1982,1984). Moreover, as will be further discussed in a later section, although a review of very recent literature will indicate that a great many tropical fruit species have been successfully propagated *in vitro*, the number of such reports dealing with successful *in vitro* clonal micropropagation using organized tissue explants taken from mature trees is extremely low. Amin and Jaiswal (1987) have only recently reported the successful propagation *in vitro* of the important tropical fruit crop guava (*Psidium guajava*), using nodal explants taken from mature trees.

Although there are few available reports that describe successful micropropagation of tropical fruit crops using mature tissues, it is with certain members of this group, namely species within the family Rutaceae, that much early progress was made in the development of *in vitro* techniques with the potential for biotechnological application. Several early reports (e.g. Vardi *et al.*, 1975) described the first successful regeneration from protoplasts of woody species; these techniques were later refined (Vardi *et al.*, 1982). More recently Hidata and Kajiura (1988) were able to regenerate plants from protoplasts of three species of *Citrus* that had been maintained as callus cultures for two years.

Techniques of protoplast culture have also been applied to other economically important tropical tree species, such as coffee. Schopke *et al.* (1987) have reported somatic embryogenesis and successful regeneration of plantlets from protoplast cultures of *Coffea canephora* P. ex. Fr.

Using an extension of such techniques, Ohgawara *et al.* (1985) have reported successful regeneration of somatic hybrid plants obtained by protoplast fusion between *Citrus sinensis* and *Poncirus trifoliata*. The presence of both parental nuclear DNAs in the hybrid was confirmed by EcoRI (a restriction endonuclease enzyme) digest analysis of rDNA fragments. Similar work has been reported by Grasser *et al.* (1988), who used chromosome counts and malate dehydrogenase isoenzyme patterns to confirm the hybrid nature of plants regenerated by protoplast fusion between *Citrus sinensis* and *Ponoirus trifoliata*. Although techniques of 'somatic hybridization' using protoplast technology may hold great promise in the future for production of 'wide hybrids', the inherent necessity of regeneration from a callus phase, and the associated risk of genetic instability, would appear to make such techniques unsuitable for purposes of germplasm conservation. However, the use of *in vitro* culture techniques combined with methods of genetic engineering may provide a pathway for the rapid genetic improvement of woody species. Such systems could also aid in the rapid transfer of desirable genetic traits from wild relations of cultivated crop species into current commercial varieties. These wild relations are presently maintained in 'germplasm banks', and possess many undesirable attributes. The transfer of desirable characters from these species into a desirable genetic background using conventional methods of plant breeding is both extremely slow and difficult. This topic will be discussed in more detail in the next section.

Almost simultaneously with the development of reliable and efficient techniques for *in vitro* multiplication of temperate fruit-tree rootstocks, micropropagation methods were being applied to numerous other woody species. As in the case of fruit-tree rootstocks discussed above, the motive for initiating most such early work was financial, with research efforts concentrating on commercially valuable woody ornamentals such as rhododendrons (Anderson, 1975; Kyte and Briggs, 1979) and roses (Hasegawa, 1979). Other researchers active during this period centred their efforts on problems of propagation affecting the major commercial palms. Clonal plantlets of date palm were first obtained by several workers, including Reynolds and Murashige (1979). These initial reports were successfully followed up by Tisserat (1979,1982). Progress with oil palm was initially slow, but two teams of workers, in England (Jones, 1974) and in France (Rabechault and Martin, 1976), eventually obtained clonal plantlets. Since then, the technique has undergone considerable development (Ahee *et al.*, 1981; Pannetier *et al.*, 1981). With the coconut palm, *in vitro* plantlets were not obtained until after some ten years of intensive research effort (Branton and Blake, 1983). Blake (1983) has reviewed the development and application of tissue-culture techniques to propagation of palm species. In the case of bananas and plantains, rapid early progress was made in the development of tissue-culture methods of multiplication, and systems of *in vitro* propagation have been available for some 15 years (Ma and Shii, 1972,1974).

8.3.2 Applications of tree tissue culture

(a) Forest trees

(i) GYMNOSPERMS The *in vitro* culture of conifers has been reviewed extensively in recent years (Mott, 1981; Biondi and Thorpe, 1982; David, 1982; John, 1983; Thorpe and Biondi, 1984; Bonga, 1987; Bonga and Durzan, 1987). Many coniferous species have now been introduced into culture, and explants used for culture initiation have been derived from almost every tissue present during the life history of conifers, including: male and female gametophytic tissue; immature and mature embryos; hypocotyls, cotyledons, and shoot apices of seedlings; shoot apical meristems; shoot tips; juvenile and mature needles; brachyblasts; and cambial tissues. However, as Bonga (1987) has very recently pointed out, successful *in vitro* propagation of sexually mature Gymnosperm trees has been reported only for *Sequoia sempervirens*. In explants from mature trees of other species, only the differentiation of adventitious needles or more or less stunted adventitious shoots without roots were obtained.

Evers *et al.* (1988) have recently reported a study aimed at micropropagation of Douglas fir (*Pseudotsuga monziesii*). The production of superior seed of this species requires the establishment of extensive seed orchards, since trees flower abundantly only once every seven years. Attempts to use tissue-culture methods for establishment of seed orchards have so far been unsuccessful because of difficulties encountered during the process of *in vitro* root induction. Evers *et al.* (1988) found that shoot cultures derived from trees more than three years old failed to root. This situation did not improve even after extended periods of time in *in vitro* culture (up to three years). It would therefore appear that there is an urgent need to develop reliable *in vitro* regeneration systems for tissues derived from mature Gymnosperms. This subject is likely to require the input of a great deal of research effort in the future. There is a limited number of reports regarding large-scale multiplication of coniferous species. Poissonnier *et al.* (1980) have reported the production of 30 000 plants of *Sequoia sempervirens* from 200 selected clones. A small production laboratory has been built at Roturua, New Zealand, where large-scale micropropagation of *Pinus radiata* has been attempted, although from juvenile explant material (Smith *et al.*, 1982).

Very recently, an enormous amount of interest has been shown in the application of tissue-culture techniques in conjunction with methods of genetic engineering or 'gene manipulation' for forest tree improvement, with a consequent rapid expansion of research effort in this area. Genetic engineering of conifers provides the opportunity for the introduction of novel traits into important forest species far more rapidly than is currently feasible using conventional breeding methods. Although gene-manipulation techniques have been successfully demonstrated using a wide range of agriculturally important species, the application of these techniques to forestry, and in particular to conifers, has been hindered by a lack of transformation methods and efficient *in vitro* regeneration systems.

Reports by Bekkaoui *et al.* (1987) and Attree *et al.* (1987) demonstrated that protoplasts of the coniferous species *Picea glauca* (white spruce) could be regenerated to form callus and somatic embryos respectively. Very recently, Attree *et al.* (1989) have extended this work and have achieved regeneration of plantlets from embryogenic protoplasts of *Picea glauca*. Similarly, Finer *et al.* (1989) reported somatic embryogenesis from protoplasts of eastern white pine (*Pinus strobus* L.), demonstrating that such techniques are applicable to other Gymnosperm genera.

These encouraging developments indicate that *in vitro* systems of regeneration compatible with transformation techniques are feasible for coniferous species.

Concurrent with these recent advances in the development of effective *in vitro* regeneration strategies for protoplast-culture systems, several groups of research workers have investigated methods for introducing DNA into cell cultures of coniferous species, with the eventual aim of producing transgenic plants. The most widely used system of gene transfer to date has utilized the naturally occurring bacterium *Agrobacterium tumefaciens* as a vector agent. This soil-dwelling organism causes crown gall disease in many dicotyledonous plants, by inserting a segment of DNA from one of its plasmids into plant chromosomal DNA. This segment, T-DNA, contains genes that interfere with auxin and cytokinin synthesis, and this results in tumerous growth of the infected tissue. Until recently it was thought that the host range of *Agrobacterium* was mainly confined to dicotyledonous Angiosperms. However, Ellis *et al.* (1989) have successfully transformed seedlings of white spruce and other conifer species using *A. tumefaciens*, and confirmed stable integration of T-DNA-encoded genes into the host-plant genome.

Other methods of gene transfer have also been investigated, as alternatives to the use of vectors. One such method is electroporation. In this procedure, which works well with a wide variety of species of both monocotyledonous and dicotyledonous plants, an electric current is used to make temporary pores in the protoplast membrane, allowing DNA from the surrounding solution to enter the cell. High frequencies of transformation can be achieved with this method. Bekkaoui *et al.* (1988) have recently used this technique to achieve transient gene expression in protoplast cultures of *Picea glauca*.

The general topics of applying genetic-engineering techniques to woody plants and the uses of biotechnology in agriculture and forestry have been reviewed and discussed by various authors, e.g. Hanover and Keathley (1988) and Bajaj (1986).

Genetic engineering may also prove to be of great value to plant genetic conservationists, assisting in the utilization of valuable germplasm in breeding programmes. In addition to the use of molecular markers for germplasm evaluation and utilization, and as methods for monitoring the stability of stored cultures *in vitro*, recombinant DNA methodology allows the identification and subsequent transfer of a selected gene (or genes) from one plant to another. This process, as discussed earlier, could greatly speed up breeding programmes, and may enhance the use of wild relatives of crop species held in germplasm collections. This is especially pertinent in view of the fact that there is presently a growing perception that genetic resources collections are generally undervalued and under-used (Holden, 1984; Peters and Williams, 1984; Marshall, 1989). This reluctance to utilize wild relatives may be due to the immense amount of time and effort needed to transfer even a single useful character into a desirable genetic background, a problem which may be solved or considerably eased by the use of genetic-engineering techniques. Peacock (1989) has reviewed the impact that recent advances in recombinant DNA methodology are having in the area of plant genetic resources utilization.

The *in vitro* culture of woody species has also found a niche in plant-pathology studies and investigations of host-pathogen interactions. Lopez and Navarro (1981) detected the bacterium *Xanthomonas campestris*, which is responsible for bacterial canker in *Citrus* spp., by inoculating aseptically grown seedlings with a suspension of the pathogen. More recently, Abdul Rahman *et al.* (1986) investigated the interaction between the fungal pathogen *Gremmeniella abiertina* and species of *Larix* (larch), which comprise the natural host species. Under natural (i.e. *in vivo*) conditions, the pathogen normally requires one to one and a half years to complete its life cycle, but this is reduced to only four weeks under *in vitro* culture conditions.

(ii) ANGIOSPERMS In recent years, several reports have been published describing the successful *in vitro* clonal propagation of mature forest trees. Gupta *et al.* (1979,1981) have described the successful multiplication *in vitro* of elite trees of both teak (*Tectonia grandis*) and *Eucalyptus citriodora*. Selection criteria for the former species were based on wood and grain quality, whilst elite trees of the latter species were identified on the basis of qualities such as disease resistance and leaf-oil content. Shortly after, a report by Datta *et al.* (1982) described the *in vitro* propagation of the valuable tropical timber *Dalbergia sisso* (one of the rosewoods). A later report by Lakshmi-Sita *et al.* (1986) described plant regeneration from

179

shoot-derived callus of *Dalbergia latifolia*, the most important rosewood species of India. Additional reports by various workers have described the successful *in vitro* culture of several other valuable timber species, including *Fragraea fragrans*, a slow-growing species which is indigenous to south-east Asia, and whose wood provides a valuable timber of the heavy hardwood group (Lee and Rao, 1986). Similarly, Paily and D'Souza (1986) have recently reported the tissue-culture propagation of *Lagerstroemia flos-reginae*, using explants taken from 30-year-old trees. This deciduous species is considered to be one of the most beautiful trees of India, and is valued for its flowers, timber and medicinal properties. In addition, the species has considerable resistance to moisture stress, and has good potential for reforestation of degraded areas. However, previous attempts at large-scale multiplication have proved difficult, due to very poor seed germination.

In addition to the many tropical timber species, attention has also been given to the *in vitro* clonal propagation of important temperate hardwood species. Two recent reports by Nepveu (1982,1984) have shown that certain characters associated with desirable wood quality in two species of oak (*Quercus robur* and *Q. petraea*) are genetically determined and hence would be lost through genetic recombination in seed-propagated offspring. Favre and Juncker (1987) recently achieved *in vitro* growth of buds taken from stump sprouts of a 150-year-old forest tree selected for its wood quality. This latter report highlights the tendency of some workers, when attempting to establish shoot cultures *in vitro* from mature trees, to select tissue explants from zones or regions of an older (mature) tree which still possess a physiologically juvenile status and hence are more amenable to *in vitro* culture.

A recent volume by Evers *et al.* (1988) gives a comprehensive summary of research carried out at the Physiology Department of the Dorschkamp Research Institute for Forestry and Landscape Planning, Wageningen, Netherlands, between 1983 and 1986, and dealing with aspects of the tissue culture of various forest tree genera. Genera investigated included: *Alnus* (alder), *Salix* (willow), *Platanus* (plane), *Quercus* (oak), *Populus* (poplar) and *Ulmus* (elm). Techniques are described for bud, embryo, meristem and callus culture. It was found that with some genera, such as *Alnus*, it proved extremely difficult or impossible to root *in vitro* shoots derived from older trees, and the technique of grafting *in vitro* meristems onto decapitated seedlings was employed. This method results in adult (mature) meristems on juvenile root systems. Another problem encountered was the observation that, in species such as plane and oak, the topophysical variation (relating to the position on the parent tree of the buds/shoots used as explant sources, i.e. terminal/axillary and upper/lower branches) was responsible for a major part of the variation in morphogenesis *in vitro*.

With shoots of oak (*Quercus*) in the initial stages of culture, these differences in morphogenetic response were extreme. The potential for growth from buds in the upper part of the crown was 2–20 times as great as in buds in the lowest branches. The same extreme difference in growth was also found between cultures derived from terminal and axillary buds.

Another example of the successful tissue-culture propagation from mature specimens of economically important tree species concerns the report by Sutter and Barker (1985) describing the multiplication *in vitro* of shoot explants from mature *Liquidambar styraciflua* (American sweetgum). This species is an important ornamental tree of many American streets.

Various groups of workers have described the successful *in vitro* multiplication of multipurpose leguminous species, e.g. *Leucaena leucocephala* (Goyal *et al.*, 1985), and *Casuarina equisetifolia* (Duhoux *et al.*, 1986). The latter report describes shoot multiplication through induction of axillary buds on immature inflorescences, the objective being to propagate trees selected on the basis of fast growth and efficient nitrogen-fixing ability. Propagation of *Casuarina equisetifolia* has previously proved difficult using conventional methods, since rooted cuttings display plagiotropic growth.

In common with coniferous forest species, there has very recently been an upsurge of interest in the development and aplication of cell-culture techniques to deciduous forest species, with the eventual aim of developing systems for genetic engineering. Russell and McCown (1988) have reported the successful recovery of plants from protoplasts of hybrid poplar and aspen clones. Lang and Kohlenbach (1988) have reported regeneration of callus from mesophyll protoplasts of *Fagus sylvaticus* (European beech). More recently, Cheema (1989) achieved successful somatic embryogenesis and plant regeneration from cell suspensions derived from a 40-year-old specimen of *Poplus ciliata* (Himalayan poplar). Concurrent with this progress in the development of *in vitro* regeneration systems, several successes have been reported in gene-transfer investigations utilizing broadleaf forest genera. Mackay *et al.* (1988) used *Agrobacterium tumefaciens* to demonstrate genetic transformation of nine *in vitro* clones of *Alnus* (alder) and *Betula* (birch). Sequin and Lalonde (1988) have used the alternative technique of electroporation to achieve gene transfer in *Alnus incana*. The genus *Alnus* is of great economic value as a tree for forestry programmes such as land reclamation and reforestation, since members of the genus are capable of fixing atmospheric nitrogen through a symbiotic association with the actinomycete *Frankia*. There is currently great interest in the possibility of transferring plant symbiotic genes into non-nitrogen-fixing woody species.

Another interesting application of *in vitro* techniques is the prospect of creating artificial or somatic seeds. In this process, somatic embryos (usually) are encapsulated in a gel, which is then solidified and coated with a rigid shell, before storage and/or sowing. The technique has been discussed by Redenbaugh *et al.* (1986). More recently, Bapat and Rao (1988) have applied the technique to the valuable timber species *Santalium album* (sandalwood). Embryogenic cell suspensions were initiated and cultured using a 20-year-old specimen of *S. album* as explant source. Somatic embryos were encapsulated in an alginate matrix and could be stored at 4°C for 45 days before reculturing to form plantlets.

(b) Palms and plantation crops

With the exception of the coconut palm (*Elaeis guineensis*), much progress has been made concerning the application of methods of *in vitro* propagation to palm species. Within three years of the initial report by Jones (1974) describing tissue-culture propagation of oil-palm (*Elaeis guineensis*), some 20 ha of *in vitro* propagated oil palms, including over 30 clones, were planted and undergoing field trials in Malaysia (Corley *et al.* 1977,1979). Later work by Nwankwo and Krikorian (1983) demonstrated the use of tissue-culture methods for propagation of female-infertile *pisifera* oil palms, which are of value as desirable pollen sources for production of commercial *tenera* varieties. More recently, the *in vitro* oil palm propagation process developed by ORMSTOM/IRHO in France will soon be applied on an industrial scale (Noiret *et al.*, 1985). Tisserat (1984) has reviewed recent developments concerning tissue-culture propagation of the date palm (*Phoenix dactylifera*).

Following the early reports of Ma and Shii (1972,1974), which showed tissue-culture propagation of banana to be feasible, many reports appeared in the literature describing efficient methods for *in vitro* multiplication of *Musa* spp. and, as a result, *in vitro* propagation is now the norm. An early report by Berg and Bustamante (1974) described the use of meristem-culture techniques and thermotherapy for production of virus-free banana plants. Later reports by other workers (e.g. Dore-Swamy *et al.*, 1983) were concerned with the development of methods of rapid clonal propagation. More recently, Cronauer and Krikorian (1984) devised a method for rapid clonal multiplication of the plantain clones Pelipita and Saba, both of which possess high tolerance to Black Sigatoke disease, caused by the fungus *Mycosphaerella fijiensis* var. *difformis*, which is currently threatening worldwide banana and plantain production. Gupta (1986) has described a technique of bud induction on excised meristems for the eradication of mosaic disease and the rapid clonal multiplication of both bananas and plantains. Jarret (1986) has recently reviewed all aspects of the *in vitro* propagation and genetic conservation of bananas and plantains.

(c) Temperate fruits

In the ten years that have elapsed since Jones *et al.* (1977) described the rapid multiplication of virus-indexed apple rootstocks, many workers have applied the technique of rapid *in vitro* clonal micropropagation to a wide range of woody temperate fruit species and genera. These have included: self-rooted difficult-to-propagate apple scion varieties (Jones *et al.*, 1979); crabapple cultivars such as *Malus sieboldii* cv. Zumi (Singha, 1982); various pear varieties (Lane, 1979; Cheng, 1980; Singha, 1980); scion cultivars of plum (Rosati *et al.*, 1980), almond (Rugini and Verma, 1982) and sweet cherry (Snir, 1982); peach (Miller *et al.*, 1982); walnut rootstocks (Driver and Kuniyuki, 1984); woody cane fruits such as blackberry, blueberry (Zimmerman, 1978; Broome and Zimmerman, 1978) and raspberry (Anderson, 1980); and woody vines such as grape, reviewed recently by Krul and Mowbray (1984). Other reviews of temperate woody fruit plant micropropagation have been given by various workers, both general (Lane, 1982; Wilkins and Dodds, 1983b) and with reference to specific crops such as apple (Zimmerman, 1984).

With the exception of perhaps the palms and certain ornamentals, most commercial applications of woody-plant micropropagation techniques over the last decade have concerned the horticulturally important temperate fruit crops. Almost immediately after the publication of the first reports involving the experimental micropropagation of fruit-tree rootstocks, several commercial ventures were initiated to provide large-scale production of virus-free plants of the dwarfing apple rootstock M27, and the plum rootstock Pixy (*Prunus insititia*) (Jones, 1979). In Italy, a large-scale production of apple rootstocks by micropropagation was underway by 1979 (Zimmerman, 1979), and within three years similar techniques were being applied to peach rootstocks, nearly nine million plants having been produced by mid-1982 (Loreti and Morini, 1982). More recently, cultivars of various fruit trees (apple, peach, plum) growing on their own roots have been produced in limited quantities by several commercial laboratories for evaluation of the usefulness of self-rooted cultivars (Zimmerman, 1985).

As in the case of forest species, temperate fruit genera, especially members of the family Rosaceae, have been the subject of several very recent investigations aimed at the development of *in vitro* regeneration systems and also methods for gene transfer, with the eventual aim of producing genetically engineered fruit trees. Ochatt *et al.* (1987) have demonstrated whole-plant regeneration from protoplasts of the cherry rootstock Colt (*Prunus avium* × *P. pseudocerasus*). These techniques were later refined (Ochatt *et al.*, 1988). Similar methods have also been applied to other temperate fruits, but with limited success. Doughty and Power (1988) demonstrated callus regeneration from leaf protoplasts of the apple scion cultivar Greensleeves (*Malus domestica* Borkh.). Very recently, James *et al.* (1989) have shown that genetic transformation of apple is feasible using a vector system comprising *Agrobacterium tumefaciens* carrying a disarmed and modified Ti plasmid. Infection was achieved using portions of leaves excised from *in vitro* shoot cultures. Transformed shoots were regenerated directly from these explants, thereby obviating the necessity for a protoplast-culture phase.

(d) Ornamental species

Immediately following the reports of Anderson (1975,1978a,1978b) concerning experimental micropropagation of rhododendron species, several commercial growers began to utilize such techniques for mass propagation. These commercial growers have since gone on to refine and modify the original methodologies devised by Anderson, and annual production of numerous varieties and species of rhododendron and azalea is currently of the order of one million plants (Zimmerman, 1985). Similarly, research on *in vitro* propagation of roses began simultaneously in various laboratories (Hasegawa, 1980; Martin *et al.*, 1981; Skirvin and Chu, 1979), and by 1982 a large French nursery had commenced large-scale production (Delbard, 1982). Annual production by 1983 was approximately 500 000 plants (Zimmerman, 1985).

8.3.2 Tree tissue-culture methodology

Many workers have discussed the experimental methodologies involved in the propagation of woody species through *in vitro* systems. Several reviews have been compiled giving detailed experimental techniques for the efficient micropropagation of different groups of tree crops, such as the temperate fruits (Jones, 1979; Skirvin, 1981; Hutchinson, 1982; Wilkins and Dodds, 1983b; Zimmerman, 1984; Wilkins *et al.*, 1985), palms (Blake, 1983; Tisserat, 1984), bananas and plantains (Jarret, 1986), and forest species (Bonga and Durzan 1982b; Bonga and Durzan, 1987). A recent volume by Bonga and Durzan (1987), consisting of 30 specialist chapters by various authors, deals with the *in vitro* propagation of a wide variety of forest/plantation species, including Angiosperm monocotyledonous and dicotyledonous species, and also Gymnosperms.

Just as conventional methods of tree propagation vary immensely with the species or type of crop under consideration, so tissue-culture systems of propagation vary according to the category of woody crop species (i.e. whether temperate deciduous fruit, broadleaf forest species, Angiosperm monocotyledonous palm or Gymnosperm) under investigation. Although it is not the purpose of this chapter to provide extensive experimental details of *in vitro* tree-propagation systems, it is pertinent to discuss the major forms of morphogenetic pathways encountered in tree tissue-culture methodologies, since these features may prove of paramount importance when considering the potential utilization of such systems for purposes of tree germplasm conservation. The type of morphogenetic pathway leading to eventual plant regeneration obviously determines the type of available storage propagule (i.e. callus, single cells, embryoids, shoot apices, etc.) if such a system were to be incorporated into a method of germplasm preservation.

Rao and Lee (1986) have recently defined three patterns of *in vitro* differentiation leading to plantlet formation in tissue-culture systems applicable to woody plants and plantation crops. These are as follows:

1. Explant → axillary buds → multiple shoots → roots → plantlets.
2. Explant → callus → cells → embryoids/embryos → plantlets.
3. Explant → callus → meristemoids → shoots and roots → plantlets.

The first of these systems (the 'multiple-shoot technique') is the preferred method of *in vitro* propagation for all crop types, for reasons discussed later; it has been employed for clonal multiplication of a wide range of woody-plant species, such as temperate tree fruits and nuts, cane fruits, many forest species and some tropical fruit and timber species.

In basic terms, the technique involves the excision of a shoot-tip/meristem explant derived from either a terminal or lateral bud of the desired parent tree, followed by surface sterilization and subsequent culturing on a defined mineral salt medium containing a carbohydrate source and vitamins, and supplemented with a cytokinin. Multiple shoots are then allowed to develop through the outgrowth of preformed axillary meristems and, once established, proliferating shoot cultures may theoretically be maintained indefinitely, or at least until the desired number of shoots have been obtained, by removing a shoot or clump of shoots and transferring to fresh medium of the same composition. When desired, adventitious roots may be initiated by exposure of the basal region of individual excised shoots to a medium in which the cytokinin has been omitted and an auxin included. Once roots have been initiated, shoots may be transferred to a medium devoid of growth regulators to allow root elongation. When a satisfactory root system is obtained, plantlets are usually transferred to a potting mixture and grown on in a greenhouse under high humidity and with suitable illumination, until established *in vivo*.

In contrast to the multiple-shoot technique just described, in which organized culture growth proceeds at all stages throughout the propagation system, the other two types of *in vitro* culture system listed earlier include a disorganized callus phase, and may also involve the initial induction of callus on a tissue explant consisting of a non-meristematic plant part, such as a portion of leaf or a stem or root segment. The initial callus-induction phase is usually achieved through culturing on a defined medium containing a high concentration of an auxin such as naphthalene acetic acid (NAA) or 2,4-dichlorophenoxyacetic acid (2,4-D). Once active callus proliferation and growth have been attained, morphogenesis (bud or embryoid differentiation) may be induced by manipulation of the types and concentrations of growth regulators within the culture medium. If callus differentiation results in bud formation, then individual expanded shoots may be subsequently excised, rooted as described for the multiple-shoot technique, and then transferred to *in vivo* conditions. In the case of morphogenesis resulting in embryoid development, individual embryos are subsequently transferred to fresh nutrient medium to continue development prior to transfer to *in vivo* conditions. Such a propagation system may be maintained, actively producing plants for long periods, by regular subculturing of the original morphogenetic callus line to medium that promotes production of new morphogenetic callus.

Species such as the monocotyledonous date palm (*Phoenix dactylifera*) and oil palm (*Elaeis guineensis*), in which the branching habit is either infrequent or absent, are presently propagated by such means, although the methods employed differ slightly according to the original source of explant material used. Engelmann *et al.* (1987) have described the propagation, on an industrial scale, of oil palm plants via somatic embryogenesis in leaf-derived callus. However, a different approach was employed by Tisserat *et al.* (1981), who developed a system of tissue-culture propagation of date palm by somatic embryogenesis in callus cultures derived from shoot-tip explants (i.e. meristematic tissue) dissected from clonal offshoots. With arborescent monocotyledons such as the palms, there is an obvious reluctance to use terminal meristematic tissue for purposes of initiating cultures, since dissection of this region of tissue from the plant kills the palm. Other workers attempting to propagate palms *in vitro* have utilized immature inflorescence tissue as a source material for induction or morphogenetic callus cultures.

Unfortunately, an inherent drawback of tissue-culture systems involving a disorganized callus phase is that subsequent development of shoots or embryos is via an adventitious pathway, and as such is usually considered to be genetically unstable. Many workers have reviewed the problems of genetic instability in plant tissue cultures in recent years (D'Amato, 1975,1978; Bayliss, 1980; Reisch, 1984; Scowcroft, 1984).

8.3.3 Tissue-culture conservation of woody species

Great progress has been made in the field of plant genetic conservation during the 20 years that have elapsed since the joint FAO/IBP technical conference was held in Rome during 1967 (Bennet, 1968) to discuss the status (at that time) of plant genetic resources exploration, evaluation, utilization and conservation. A global network of 'gene banks' or seed-storage facilities has since been established within International Agricultural Research Centres (IARCs) in many countries throughout the world. Examples of such IARCs include the International Rice Research Institute (IRRI) in the Philippines, which maintains the world collection of rice germplasm, and also Centro Internacional de Mejoramiento de Maiz y Trigo (CIMMYT) in Mexico, which holds world collections of maize and wheat germplasm.

With most annual, inbreeding, seed-propagated crops such as rice, maize and wheat, the technical problems of genetic conservation were solved fairly readily, because the seed of these species (and most similar crop types) is classified as 'orthodox' and may be maintained under conditions of low moisture content (5%) and temperature ($-20°C$). However, such methods were not applicable to clonally propagated outbreeding crops or species whose seed is termed 'recalcitrant' and hence cannot be stored by conventional techniques. Initially, germplasm of these crops was stored vegetatively, with (for example) the world collection of potato (*Solanum tuberosum*) being maintained as annually cultivated field collections at Centro Internacional de la Papa (CIP) in Peru, whilst vegetative material of cassava (*Manihot esculentum*) was maintained at Centro Internacional de Agricultura Tropical (CIAT) in Colombia. Subsequently, the rapid development of tissue-culture methods of propagation for herbaceous species such as potato, which has proved very amenable to *in vitro* culture techniques, has provided a method for clonal storage of potato germplasm under aseptic conditions, free from the risk of losses due to pathogen attack, to which vegetatively maintained collections are prone. Subsequently, tissue-culture methods were developed for the tropical starchy staple crop cassava. The conservation *in vitro* of both potato and cassava is dealt with in detail in Chapters 5 and 6.

With the recent rapid expansion in application of tree tissue-culture methods, and the concurrent growing worldwide awareness of the need for conservation of many species of endangered woody plants, many scientists have speculated that *in vitro* methods may provide an effective method for germplasm conservation of these crops. However, progress in this area to date has been extremely slow. A recent extensive bibliography of plant tissue-culture literature compiled by Bhojwani *et al.* (1986) lists some 174 references dealing with the application of tissue-culture techniques to plant germplasm storage and includes over 60 repetitive review articles and chapters. Out of the original total, however, only 12 reports concern *in vitro* storage of woody plants and trees, of which five are reviews, four involve storage of non-morphogenetic culture systems (i.e. cell suspensions) and only five describe the *in vitro* storage of organized cultures (i.e. shoots/plantlets) of woody species, or, alternatively, non-

Table 8.1 Genera of woody plants investigated, or under investigation with a view to *in vitro* storage (information from IBPGR *in vitro* Conservation Database: April, 1987)

Acer	*Cryptomeria*	*Mangifera*	*Quercus*
Actinidia	*Dalbergia*	*Morus*	*Rhododendron*
Aesculus	*Daphne*	*Musa*	*Ribes*
Anacardium	*Disanthus*	*Olea*	*Rosa*
Araucaria	*Elaeis*	*Perilla*	*Rubus*
Artocarpus	*Eucalyptus*	*Phoenix*	*Schizophragma*
Boronia	*Eugenia*	*Pinus*	*Syringa*
Carica	*Hevea*	*Pistacea*	*Tectonia*
Castanea	*Juglans*	*Poncirus*	*Theobroma*
Chamaepericlymenum	*Larix*	*Populus*	*Tilia*
Cinnamomum	*Liquidambar*	*Prunus*	*Vaccinium*
Citrus	*Magnolia*	*Pseudotsuga*	*Viburnum*
Cocos	*Mallotus*	*Punica*	*Vitis*
Coffea	*Malus*	*Pyrus*	

organized but morphogenetic callus cultures. In spite of the disappointingly low number of relevant reports cited by the aforementioned bibliography, it is encouraging to note that a recent search (conducted April, 1987) of the IBPGR *in vitro* database (dealing with current research concerning *in vitro* propagation and storage of plant genetic resources: for details see Withers and Wheelans, 1986) who showed that 197 out of the total of 1341 available records were concerned with *in vitro* storage of woody species. The survey indicated that approximately 56 genera of woody plants were presently under study with a view to eventual development of *in vitro* storage techniques (see Table 8.1).

8.3.4 Methods of *in vitro* storage

Many workers in recent years have reviewed the various possible experimental approaches to storage of plant cell, tissue and organ cultures by *in vitro* techniques, and the advantages inherent therein (Withers, 1980,1984; Wilkins and Dodds, 1983a,c; de Langhe, 1984; Kartha, 1985). A recent review by Aitken-Christie and Singh (1987) has discussed aspects of cold storage of woody-tissue cultures, whilst Sakai (1985) has reviewed the method of cryopreservation of shoot tips as a means of germplasm conservation for fruit trees. In addition, the methodologies involved in the two most commonly employed techniques of *in vitro* plant germplasm conservation, i.e. reduced-growth storage and cryopreservation, are discussed in detail in Chapters 2 and 3.

8.3.5 Applications

Examples of the successful application of slow-growth techniques for storage of woody species are given in Table 8.2, whilst available reports and information relating to cryopreservation of woody plant tissues are summarized in Table 8.3. It should be noted that the reports cited in Table 8.3 refer only to cryopreservation of organized tissues (i.e. buds, embryos, meristems), or morphogenetic cell cultures from which differentiation of organized structures such as shoots or embryos was subsequently induced after storage. There are several additional reports available that deal with the cryopreservation of woody species (Sakai and Sugawara, 1973; Sugawara and Sakai, 1974; Withers, 1978; Pritchard et al., 1982). However, these reports were 'successful' only in the sense of maintenance of cell viability after storage, without subsequent differentiation of organized structures that would eventually lead to establishment of free-living plants. Although such examples are of value for studying aspects of the processes of intracellular freezing injury in plant tissues, these systems would have limited, if any, value for purposes of germplasm preservation. Merryman and Williams (1985) have recently reviewed various aspects related to freezing injury in plant cells.

The reports listed in Table 8.3 indicate that various workers have employed several different experimental approaches to the *in vitro* conservation of woody species using methods of cryopreservation. These include storage of: (1) zygotic/nucellar embryos, (2) dormant buds/apices, (3) somatic embryoids, (4) morphogenetic callus cultures, and (5) *in vitro* shoot apices. It will be evident that these approaches fall into two broad categories, depending on whether the tissue explants being stored are excised from *in vivo* plants or are derived from or already part of an *in vitro* clonal propagation system.

Although cryopreservation of dormant buds of woody species has been demonstrated (Sakai and Nishiyama, 1978; Stushnoff, 1987), this method has the disadvantage that survival during cryostorage is to some extent dependent on the inherent cold-hardiness of the particular cultivar under study (Stushnoff, 1987). With less hardy cultivars, it was necessary to pass buds through a defined regime of dehydration and subsequent cold-acclimation, or to select buds at a time of maximum 'cryopreservability', when the buds are at the optimum physiological stage of development to survive dehydration and subsequent cryopreservation (Stushnoff, 1987).

With palm species, two different approaches have been used during successful cryopreservation protocols; these involve either the storage of morphogenetic callus lines and subsequent regeneration of plants after thawing (Tisserat et al., 1981), or the cryopreservation of *in vitro* somatic

Table 8.2 Summary of reports dealing with the slow-growth storage of woody species

Species/cultivar	Common name or use	Duration of storage	Slow-growth method(s) used	Reference
Actinidia chinensis	Kiwi fruit	9 months	Storage at 25°C on medium containing activated charcoal	Wilkins *et al.* (1988)
Actinidia chinensis	Kiwi fruit	12 months	Reduced culture temperature (8°C)	Monette (1986)
Alnus cordata	Italian alder	Approx. 12 months	Various	Barghchi (1987c)
Alnus glutinosa	Alder	12 months	Reduced culture temperature (6°C) and low light intensity	Evers *et al.* (1988)
Castanea sativa	Chestnut	17 months	Reduced culture temperature (2–3°C)	A. M. Vieitez (personal communication)
Cinchona ledgeriana	Source of alkaloids (i.e. quinine)	42 days	Reduced culture temperature (12°C) and use of mannitol	Hunter (1986)
Coffea arabica cultivars Caturra Rojo and Catuai	Coffee	Over 2 years	Reduced concentration of sucrose in culture medium	Kartha *et al.* (1981)
Cyphomandra betaceae	Tamarillo	Approx. 12 months	Various	Barghchi (1987c)
Eucalyptus spp.	Eucalyptus	Not stated	Reduced culture temperature (4–6°C)	Aitken-Christie and Singh (1987)
Eucalyptus tereticornis E. torelliana	Eucalyptus	Not stated	Reduced culture temperature (15°C)	A. F. Mascarenhas (personal communication)
Liquidambar styraciflua	American sweetgum	7 months	Reduced culture temperature (4°C)	T. R. Marks (personal communication)
Malus baccata	Ornamental crabapple	18 months	Reduced culture temperature (4°C)	Wilkins *et al.* (1988)

190

Species	Description	Duration	Method	Reference
M. domestica, cultivars: (i) M7, M9, M25, M26, M27 (ii) Greensleeves, Wijcik	(i) Domestic apple rootstocks, dwarfing and invigorating (ii) Domestic apple scions	12–28 months (dependent on cultivar and storage method)	Various, including: reduced culture temperature, various growth retardants, mannitol, increased volumes of medium and use of activated charcoal	Wilkins *et al.* (1988)
M. domestica cv. Golden Delicious	Domestic apple scion	12 months	Reduced culture temperature (1°C and 4°C)	Lundergan and Janick (1979)
M. prunifolia	Rootstock/ornamental	18 months	Reduced culture temperature (4°C)	Wilkins *et al.* (1988)
Morus nigra	Black mulberry	9 months	Storage at 25°C on medium containing activated charcoal	Wilkins *et al.* (1988)
Musa spp. (eight triploid cultivars including genotypes: AAA, AAB and ABB)	Bananas and plantains	13–17 months	Reduced culture temperature (15°C)	Banerjee and de Langhe (1985)
Pinus radiata	Radiata pine	5.5 years (with one subculture)	Reduced culture temperature (4–5°C)	Aitken-Christie and Singh (1987)
Pistacia vera	Pistachio nut	Almost one year	Various	Barghchi (1987c)
P. mutica *P. khinjuk* *P. atlantica*	Pistachio rootstocks	Not stated	Not stated	M. Barghchi (personal communication)
Populus alba × *P. glandulosa*	Poplar	12 months	Reduced culture temperature (4°C)	Evers *et al.* (1988)
P. trichocarpa (three clones)	Poplar	12 months	Reduced culture temperature (4°C)	Evers *et al.* (1988)

Table 8.2 *Continued.*

Species/cultivar	Common name or use	Duration of storage	Slow-growth method(s) used	Reference
Prunus avium × *pseudocerasus* cv. Colt	Semi-dwarfing sherry rootstock	12–15 months	Various	Wilkins *et al.* (1988)
P. cerasus cv. CAB IIE	Sweet cherry rootstock	10 months	Reduced culture temperature (−3°C and 4°C)	Marino *et al.* (1985)
P. cerasifera cv. pissardii	Ornamental plum	12 months	Reduced culture temperature (4°C)	Wilkins *et al.* (1988)
P. domestica cv. Jeffersons Gage	Domestic plum scion	12–18 months	Various	Wilkins *et al.* (1988)
P. domestica cv. D1869	Peach rootstock	10 months	Reduced culture temperature (−3°C and 4°C)	Marino *et al.* (1985)
P. insititia cv. Pixy	Plum rootstock	12–18 months	Various	Wilkins *et al.* (1988)
P. padus	Bird-cherry (ornamental)	12–15 months	Various	Wilkins *et al.* (1988)
P. persica × *P. amygdalus*	Peach/almond hybrid	10 months	Reduced culture temperature (−3°C and 4°C)	Marino *et al.* (1985)
Punica granatum	Pomegranate	18 months	Reduced culture temperature (10°C)	Wilkins *et al.* (1988)
Pyrus communis cv. Caucasica	Pear	Up to 18 months	Reduced culture temperature (4°C)	Wanas *et al.* (1986)
P. pashia	Wild pear	12 months	Reduced culture temperature (4°C and 10°C)	Wilkins *et al.* (1988)
Ribes spp.	Red currants and blackcurrants and gooseberries	30 months	Reduced culture temperature (1–4°C)	J. F. Seelye (personal communication)

192

Species	Common name	Duration	Treatment	Reference
Robinia pseudoacacia	Black locust	Almost one year	Various	Barghchi (1987c)
Rosa (various cultivars)	Ornamental rose cultivars/hybrids	Up to 12 months	Reduced culture temperature (11°C)	A. R. Rosten (personal communication)
Rubus idaeus (64 cultivars/selections)	Red raspberry			
Rubus hybrids (46 cultivars/selections)	Bramble cultivars	30 months	Reduced culture temperature	J. F. Seelye
Rubus hybrids (5 cultivars/selections)	Purple raspberry		(1–4°C)	(personal communication)
R. occidentalis (2 cultivars)	Black cap raspberry			
Tectonia grandis	Teak	Not stated	Reduced culture temperature (15°C)	A. F. Mascarenhas (personal communication)
Ulmus spp.	Elm	12 months	Reduced culture temperature (4°C)	Evers *et al.* (1988)
Vaccinium spp.	Blueberry	30 months	Reduced culture temperature (1–4°C)	J. F. Seelye (personal communication)
Vitis spp. (7 species and 3 hybrids)	Grape	6–12 months	Reduced culture temperature (9.5°C)	Barlass and Skene (1983)
Vitis spp.	Grape	6 and 12 months	Reduced culture temperature (9.5°C)	Galzy (1969)

Table 8.3 Summary of reports dealing with cryopreservation of woody species

Species/cultivar	Common name or use	Plant part or type of culture explant stored	Type of regeneration after cryopreservation	Reference
Araucaria hunsteinii	Klinkii pine	Zygotic embryos	Survival of root meristem tissue and subsequent production of callus	Grout (1986) Pritchard and Prendergast (1986)
Citrus spp.	Citrus fruit	Ovule	Regeneration of shoots after thawing and subsequent establishment of plants *in vivo*	Bajaj (1984)
Citrus sinensis cv. Washington navel	Orange	Nucellar embryos	Cultured thawed embryos produced additional embryos, pseudobulbi and plants. Plants were successfully established in greenhouse	L. Navarro (personal communication)
Cocos nucifera	Coconut palm	Zygotic embryos	Thawed explants produced callus only	Bajaj (1984)
Elaeis guineensis	Oil palm	Zygotic embryos	Thawed embryos cultured *in vitro* developed into normal plants. Field trials in progress	Grout *et al.* (1983)
Elaeis guineensis (2 clones)	Oil palm	Adventitious somatic embryoids from *in vitro* propagation system	Normal plant development	Engelmann (1985) Engelmann and Duval (1986) Engelmann *et al.* (1987)
Malus spp.	Apple	Dormant buds	Stored buds initiated new growth after grafting onto two-year-old rootstocks	Sakai and Nishiyama (1978)

Species	Common name	Material	Result	Reference
Malus spp., cv. Fuji	Apple scion	Small shoot tips/meristems (1 mm) dissected from dormant buds of field-grown trees	Shoot tips cultured *in vitro* after thawing grew and multiplied shoots	Katano *et al.* (1983)
Malus spp., cv. Fuji	Apple scion	Shoot tips isolated from *in vitro* shoot cultures	Not stated	Katano *et al.* (1984)
Malus spp.	Apple	Shoot tips isolated from *in vitro* shoot cultures	Over 80% of shoot tips resumed normal shoot growth and proliferation after storage	Kobayashi *et al.* (1987)
Malus spp., cv. Fuji	Apple	Shoot tips isolated from *in vitro* shoot cultures	Shoots resumed normal growth after storage	Ishihara *et al.* (1987)
Malus spp. (several cultivars)	Apple	Dormant buds	Cryopreserved buds patch-budded to rootstocks after storage. Flowering after 18–24 months. (Bud survival dependent on hardiness of cultivar)	Stushnoff (1987)
Phoenix dactylifera cv. Medjool	Date palm	Embryogenic callus derived from lateral buds dissected from clonal five-year-old offshoots	Regeneration of plantlets from cryopreserved callus	Tisserat *et al.* (1981)
			Analysis of isoenzymes used to assess stability of plantlets after storage	Ulrich *et al.* (1982)

Table 8.3 *Continued*

Species/cultivar	Common name or use	Plant part or type of culture explant stored	Type of regeneration after cryopreservation	Reference
Pica glauca	White spruce	Cell suspension	Regeneration of plantlets and shoots from calli regrown from cryopreserved cells maintained for one year in liquid nitrogen	Kartha *et al.* (1988)
Rubus leucodermis R. spectabilis R. idaeus* cv. Heritage *Rubus* spp., cultivars Logan Thornless and Merton Thornless	Berry fruits	Meristems isolated from *in vitro* shoot cultures	Organized apical growth after storage	Reed and Lagerstedt (1987)
Ulmus americanus	American elm	Shoot-regenerating callus line	Regeneration of plantlets from cryopreserved callus	Ulrich *et al.* (1984)

'embryoids' regenerated from a tissue-culture propagation system (Engelmann *et al.*, 1987). The former study involved successful regeneration of plantlets from embryogenic date palm (*Phoenix dactylifora* L. var. Modjool) callus cultures cryopreserved by initial slow freezing to −30°C in a programmed freezer, prior to transferring to liquid nitrogen for storage. The embryogenic callus lines utilized were initiated originally from lateral buds dissected from five-year-old offshoots. After rapid thawing at 40°C, calli were transferred to fresh nutrient medium and plantlets developed with distinct root and shoot apices after a few subsequent weeks in culture.

To date, there has been only one report of the successful cryopreservation of isolated shoot apices excised from mature trees (Katano *et al.*, 1983). However, such a method has the inherent disadvantage that shoot tips must be cultured *in vitro* after cryostorage to allow subsequent shoot growth/multiplication, and regeneration of plants. The initial use of *in vivo* material also carries an enhanced risk of microbial contamination when apices are subsequently transferred to aseptic conditions. This method also assumes that techniques of *in vitro* rooting and plantlet establishment are already available, to allow eventual regeneration of free-living plants. It therefore follows that a preferable system of obtaining storage propagules suitable for cryopreservation would be initially to establish shoot apices of the desired tree or woody plant *in vitro*. This procedure would also allow for the use of larger shoot tips to facilitate culture establishment from elite mature trees, which would be an essential feature of any system of *in vitro* tree storage. Once proliferating shoot cultures had been established, reliable systems of adventitious root induction and *in vivo* plantlet establishment could be formulated. Subsequently, or concurrently, the established proliferating shoot cultures would serve as a source of *in vitro* shoot apices for cryopreservation investigations. This method also provides a renewable supply of storage propagules for use in studies aimed at determination of the most effective freeze–thawing protocols to ensure optimum survival. In addition, Caswell *et al.* (1986) have recently demonstrated that proliferating shoot cultures of woody species such as MM106 apple rootstock (*Malus domestica*) and Smoky saskatoon (*Amelanchier alnifolia*) are amenable to *in vitro* cold-hardening techniques, a process which may enhance survival during subsequent minimal-growth or cryopreservation procedures.

Successful cryopreservation of *in vitro* shoot apices has been reported for both apple (Ishihara *et al.*, 1987) and *Rubus* spp. (Reed and Lagerstedt, 1987). However, there is an obvious need to further extend these techniques to other woody species, including other temperate deciduous tree fruits and nuts, berry fruits, forest species and the tropical hardwoods.

The technique of cryopreservation for conservation of plant germplasm resources has the advantage that storage at the temperature of liquid nitrogen ($-196°C$) involves a complete cessation of cell metabolic activity. However, the viability of material during storage cannot be assessed, and regeneration of cryostored propagules after thawing usually requires passage through an *in vitro* propagation system, to achieve subsequent shoot multiplication or plantlet regeneration. In addition, cryopreservation techniques are not yet sufficiently developed to be employed with confidence for purposes of establishment of woody-plant germplasm banks, irrespective of the species to be stored. Even in instances where the successful cryopreservation of organized tissues of woody species has been reproducibly demonstrated, percentage levels of survival relating to the resumption of normal (organized) growth and subsequent shoot/plant development after thawing have been low. For example, Engelmann *et al.* (1987) have developed a system for cryostorage of somatic embryoids of oil palm (*Elaeis guineensis*) regenerated from a tissue-culture propagation system. Various freezing protocols were investigated, and several different oil palm clones were utilized. Storage times in liquid nitrogen ranged up to 15 months. Survival of embryoids after storage ranged from 50% to 80%, but subsequent levels of resumption of normal embryoid multiplication after two months of culture *in vitro*, although reproducible, were clone-dependent and varied between only 6% and 23%. The objective of the study was to develop a system to allow the maintenance of embryoid strains produced by *in vitro* culture in order to reduce the risks of variability arising through further *in vitro* culture cycles. A report by Reed and Lagerstedt (1987) has described the successful cryopreservation of small shoot apices of five accessions of *Rubus*. Rates of resumption of shoot-apical growth after storage (although mostly high) were very variable and dependent on both the genotype and cryoprotectant treatment employed.

There have been few reports in the literature dealing with successful regeneration of woody forest species from cryopreserved tissues *in vitro*. Of the 16 references cited in Table 8.3, only three of the reports deal with cryopreservation of forest tree genera. Kartha *et al.* (1988) have recently reported the successful cryopreservation of cell-suspension cultures of *Picea glauca* (white spruce) for one year, with very high levels of viability. Calli regrown from cryopreserved cells were induced to differentiate into plantlets through embryogenesis, and into shoots by organogenesis. In view of the variable (and often very low) success rate achieved with cryopreservation systems, many workers have recommended the use of minimal-growth techniques as an alternative and more effective means of plant germplasm conservation.

The development of suitable methods of slow-growth storage for woody species often involves merely a straightforward modification of the shoot-multiplication phase during the process of *in vitro* micropropagation. Since methods of tissue-culture propagation are now well established for many genera of woody crop plants, it follows that such techniques will have been more readily applied than methods involving cryopreservation. Slow-growth methods of plant germplasm storage also possess the advantage that, although cell division still continues, albeit at a greatly reduced rate, the viability and state of health of stored culture propagules may be continuously monitored or assessed at any time during storage, by simple visual inspection.

The large number of reports listed in Table 8.2 demonstrate that a wide range of woody-plant genera/species have been successfully stored by minimal-growth techniques for periods of one year or more with maintenance of high levels of viability. It thus appears that such methods are now sufficiently developed to be employed for purposes of establishing an *in vitro* woody-plant germplasm bank.

A recent IBPGR (1986b) report has discussed all aspects of the design, planning and operation of *in vitro* gene banks. The report proposes the establishment of two distinct types of *in vitro* gene-bank facility for storage of plant germplasm resources. These comprise either an *in vitro* base gene bank (IVBG), in which plant material would be stored by cryopreservation techniques in liquid nitrogen for long periods (many tens of years), or an *in vitro* active gene bank (IVAG), in which plant material would be stored by slow-growth techniques, with periodic transfer to fresh medium at one- or two-yearly intervals, or as required. The IVAG would form part of an overall system of plant quarantine, disease indexing/eradication and maintenance/multiplication of plant genotypes under aseptic *in vitro* conditions. Such a system should also facilitate the international distribution and exchange of virus-indexed vegetative plant germplasm, in the form of *in vitro* cultures, to those countries with strict quarantine regulations. As discussed earlier in this section, IVAGs are already established and in use at some International Agricultural Research Centres, such as CIP in Peru and CIAT in Colombia, for storage of world germplasm collections of potato (*Solanum tuberosum*) and cassava (*Manihot esculentum*) respectively. It is realistic to propose that the technology now exists to establish IVAGs for conservation of many types of woody crop plants, such as the temperate fruit and nut species.

A recent study carried out by the present author at the University of Birmingham (Wilkins *et al.*, 1988) investigated minimal-growth storage of a wide range of woody fruit genera, using a variety of techniques. Shoot cultures of several cultivars of apple (*Malus* spp.), plum and cherry (*Prunus* spp.) were successfully established *in vitro* using explants taken from mature trees, whilst shoot cultures of other genotypes such as Chinese gooseberry (*Actinidia chinensis*), black mulberry (*Morus nigra*), pomegranate (*Punica granatum*) and wild pear (*Pyrus pashia*) were established using explants (shoot apices) from juvenile seedlings. Subsequently, *in vitro* systems were developed for rapid shoot proliferation, adventitious root initiation and plant regeneration. Concurrent with these studies, a broad selection of minimal-growth techniques were utilized to determine the optimum storage method that would be applicable to as wide a range of woody fruit genera as possible. These included:

1. Storage of single shoots, shoot clusters or rooted plantlets on either liquid or solid medium at reduced culture temperatures.
2. Storage of single shoots on minimal medium containing sucrose, either without growth regulators, or else supplemented with a greatly reduced concentration of either cytokinin or auxin.
3. Use of media with altered nutrient availability, or supplemented with osmotically active compounds such as mannitol, to suppress shoot growth at normal culture temperatures.
4. Use of multiplication medium supplemented with various concentrations of growth retardants, including: maleic hydrazide, chlorocholine chloride (CCC), Amo 1618, diaminazide (B9 or Alar) and abscisic acid (ABA) to suppress growth of shoot cultures at normal culture temperatures.
5. Storage of proliferating shoot cultures or rooted plantlets on increased volumes of either liquid or solid medium, to maintain viability of cultures at normal culture temperatures.
6. Storage of single shoots on multiplication medium supplemented with activated charcoal at both normal and reduced culture temperatures.

Various methods of suppressing the growth of shoot cultures at normal culture temperatures were investigated, since the commonly adopted method of reduced-temperature storage may induce cold-selection pressures over several years of storage, especially at temperatures approaching 0°C. Caswell *et al.* (1986) have already shown that the cold tolerance of shoot cultures of the apple rootstock MM106 increases by several degrees when shoots are subjected to only a ten-week *in vitro* hardening treatment.

Usually, any delay in the culturing of rapidly proliferating shoot cultures of temperature woody fruit species (normally necessary every four or five weeks if shoots are growing on agar-solidified medium) results in rapid media browning accompanied by subsequent senescence of the culture. Shoots growing on liquid medium, usually with a filter-paper 'wick' to provide support to the culture and to ensure a nutrient supply, display the phenomenon of shoot-apex blackening if subculturing is delayed more than a few weeks. A recent report by Barghchi (1987b) has indicated that the occurrence of shoot-apex blackening may be associated with both calcium and boron deficiency in the terminal meristematic region of shoots.

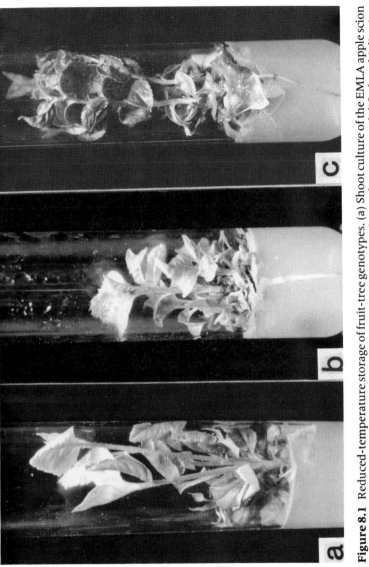

Figure 8.1 Reduced-temperature storage of fruit-tree genotypes. (a) Shoot culture of the EMLA apple scion cultivar Greensleeves (*Malus domestica*) maintained for 24 months at 4°C on agar-solidified multiplication medium. (b) Shoot culture of wild 'bird cherry' (*Prunus padus*) stored for 12 months at 4°C on agar-solidified multiplication medium. (c) Shoot culture of the EMLA plum rootstock Pixy (*Prunus insititia*) stored for 18 months at 4°C on agar-solidified multiplication medium.

It follows, therefore, that storage of shoot cultures on increased volumes of liquid culture medium should enhance culture survival at normal incubation temperatures by increasing the reservoir of available nutrients. Similarly, manipulation of the factors that influence or promote rapid shoot growth and/or proliferation (i.e. culture temperature or the concentration of endogenous growth regulators) may suppress the normal rate of culture growth and the subsequent onset of senescence. The actions of activated charcoal as an adsorption agent are well known, and inclusion in agar-solidified medium should inhibit the oxidation of polyphenolic compounds that are responsible for the onset of media browning. Alternatively, the presence of chemical growth inhibitors may sufficiently retard the rate of shoot growth and multiplication to be of value for storage purposes, whilst allowing the continued presence of the growth regulators usually essential for maintenance of the normal process of shoot multiplication.

Results obtained by this author (Wilkins *et al.*, 1988) have shown that reduced-temperature storage of various genotypes of apple, pear, plum and cherry for purposes of germplasm preservation is feasible for periods of 12 months or more at 4°C using either a liquid (some apple rootstock cultivars) or agar-based culture system. Culture viability of one apple scion cultivar (Greensleeves) was maintained at 100% for two years on agar-solidified proliferation medium (Figure 8.1a), whilst shoot cultures of five *Prunus* species were maintained at 4°C in excellent health for up to 18 months with 85% or higher levels of survival (Figure 8.1b,c).

The use of osmotic compounds (i.e. mannitol), reduced-growth regulator concentrations and various chemical growth retardants as a means of suppressing *in vitro* growth of fruit-tree shoot cultures was investigated using several temperate fruit-tree genotypes. Results obtained from all of these methods of slow-growth storage were extremely variable and cultivar-dependent. In particular, the use of reduced concentrations of growth regulators either resulted in explant non-establishment, or gave rise to a process of 'slow senescence', indicating that *in vitro* fruit-tree shoots appear to have an absolute requirement for an exogenous supply of growth regulators. Only the apple scion cultivar Greensleeves appeared amenable to growth suppression using chemical retardants (Figure 8.2), but this method was ineffective when applied to other fruit-tree genotypes, and hence these methods cannot be recommended as a suitable means of *in vitro* germplasm storage for woody species.

The use of large culture vessels containing increased volumes (50 ml) of liquid culture medium for storage of individual proliferating shoot clusters at normal culture temperatures (25°C) was moderately successful (Figure 8.3), but this method has the disadvantage that cultures easily become dislodged from the filter-paper support strip (resulting in culture senescence) if the containers are disturbed.

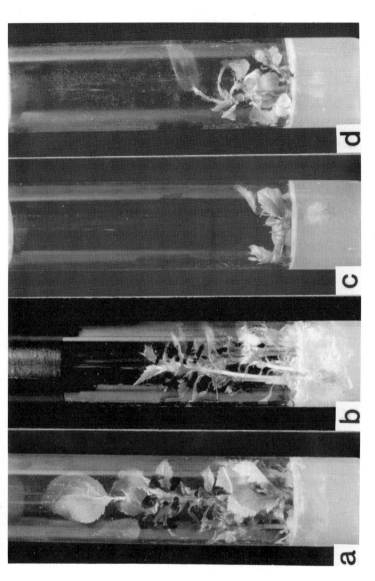

Figure 8.2 Shoot cultures of the EMLA apple scion cultivar Greensleeves (*Malus domestica*) stored *in vitro* for 12 months at 25°C on agar-solidified medium supplemented with different concentrations of various chemical growth retardants: (a) 0.1 mg/l succinic acid 2,2-dimethylhydrazine (B995); (b) 0.1 mg/l maleic hydrazide; (c) 10 mg/l Amo-1618; (d) 25.0 mg/l (2-chloroethyl)trimethyl-ammonium chloride (CCC).

Figure 8.3 Jars containing proliferating shoot cultures of two EMLA apple (*Malus domestica*) rootstock cultivars, stored for 12 months at 25°C on 50 ml of liquid multiplication medium: (a) dwarfing rootstock M7; (b) invigorating rootstock M25.

Storage of shoots at reduced temperature (4°C) on agar-solidified multiplication medium supplemented with 0.3% (w/v) activated charcoal proved to be a highly effective and widely applicable means of storage for a broad range of apple, plum and cherry species/cultivars. It was noted that no shoot proliferation occurred during reduced-temperature storage of shoots on medium supplemented with activated charcoal. This means that multiplication of stored propagules (from medium containing activated charcoal) would require an additional multiplication stage, whereas cultures stored on agar-solidified multiplication medium would provide sufficient numbers of shoots (as a result of shoot production during storage) for both restorage and utilization for other purposes, e.g. germplasm distribution, virus indexing or genetic analysis. This factor could influence the number of initial replicates being placed into storage, if such a method was adopted as part of an *in vitro* active gene-bank system.

High levels of viability of most fruit-tree genotypes stored *in vitro* on medium supplemented with activated charcoal were maintained for periods of either 12 or 15 months (Figure 8.4). Most cultures stored using this method were in an excellent state of health when storage was terminated, with no visible evidence of degeneration or senescence, and it is likely that high levels of shoot viability could be maintained for several years using this technique.

Evers *et al.* (1988) have recently reported the successful maintenance at reduced culture temperature and low light intensity of *in vitro* cultures of various woody deciduous forest species, including alder (*Alnus glutinosa*), poplar (*Poplus* spp.) and elm (*Ulmus* spp.). Rooted plantlets of alder were stored for one year at 6°C, with 70% of plantlets resuming normal growth nine weeks after returning to normal culture conditions (25°C and high light intensity). Similarly, shoot cultures of elm were stored for one year at 4°C. In the case of poplar, a system for long-distance transport of poplar cultures *in vitro* was devised. Node explants were excised from *in vitro*

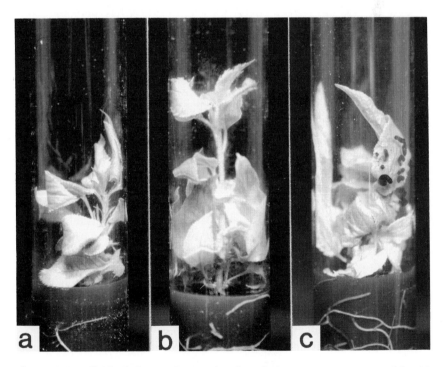

Figure 8.4 Individual shoot cultures of various fruit-tree genotypes stored for 15 months at 4°C on agar-solidified multiplication medium supplemented with activated charcoal: (a) *Malus baccata* (Siberian crab apple); (b) *M. prunifolia* (wild apple species); (c) plum scion cultivar Jeffersons Gage (*Prunus domestica*).

shoots, and placed into closed Petri dishes containing medium with 5% mannitol. Using this technique, *in vitro* cultures of poplar were sent or taken to China, New Zealand and the United States. Recovery after one week at high temperature was over 90%. Plantlets of *Poplus trichocarpa* and *P. alba* × *P. glandulosa* hybrids were maintained for one year at a reduced culture temperature of 4°C.

Results obtained from reduced-temperature storage of species such as *Actinidia chinensis* and *Punica granatum* demonstrated that storage at 4°C resulted in chilling injury and consequent rapid culture death. This observation confirms the findings of other workers (e.g. Banerjee and de Langhe, 1985) that warm-temperate and subtropical species require a storage temperature in the range of 8–15°C to avoid chilling injury. Shoot cultures of *P. granatum* have been successfully stored for 18 months at 10°C (Wilkins *et al.*, 1988), whilst Monette (1986) has recently shown 8°C to be optimal for reduced-temperature storage of shoots of *A. chinensis*. With subtropical and tropical woody crops, e.g. tropical hardwood timber species such as teak (*Tectonia grandis*), reduced-temperature storage of shoot cultures at 15°C has proved effective (A. F. Mascarenhas, personal communication). Bannerjee and de Langhe (1985) used a reduced-culture temperature of 15°C (compared to a normal culture temperature of 30°C) to successfully maintain shoots of various *Musa* cultivars *in vitro* for periods of 13–17 months. Storage of a total of eight different triploid *Musa* cultivars was investigated, including the gentoypes AAA, AAB and ABB (bananas and plantains). The authors found that the different cultivars utilized varied in their ability to withstand prolonged periods (in excess of 12 months) of minimal-growth temperatures. Cultivars with genotype AAB (plantain cultivars Asamiensa, Agbagba and Ntanga) and ABB were relatively more tolerant to reduced-temperature storage than material with the AAA genome, which displayed total loss of viability after 17 months of storage at 15°C.

Apart from the application of slow-growth storage methods for the purpose of conserving woody-plant germplasm resources, these techniques may also be incorporated into commercial systems of large-scale clonal multiplication, e.g. to allow rearrangement of subculturing dates/ times to coincide with staff availability, or to ease problems associated with variable seasonal demands. In temperate countries, the high cost of heating and lighting greenhouse facilities during winter months, usually the time also when customer demand is lowest, may mean that it is cheaper to use cold-storage methods during these periods (Suttle, 1983). The potential uses just described for *in vitro* slow-growth storage of plant material are usually short-term, with periods of storage rarely extending beyond one year.

Aitken-Christie and Singh (1987) have recently described the adoption of methods of reduced-temperature storage as part of a scheme for breeding and testing superior clones of coniferous forest species. Since 1983, a trial has been in progress at the Forest Research Institute (FRI) at Rotorua in New Zealand to concurrently cold-store and field-test 200 *Pinus radiata* clones derived from 19 of the best available FRI control-pollinated families. Fascicles of each clone were introduced into *in vitro* culture, and juvenile shoots were induced from axillary buds, then maintained *in vitro* by slow-growth storage at 4°C for a period of 5–8 years. At the same time, rooted fascicle cuttings of each clone were planted in the field, then evaluated for desirable characteristics over a similar 5–8-year period. Subsequently, once field-grown trees have reached a stage where superior clones may be selected, these elite clones may then be rapidly and easily multiplied from the 'clone-bank' collection of juvenile shoots maintained in a cold-storage facility. As discussed during the early part of this chapter, there are many obstacles related to the vegetative propagation of mature trees, with ease of propagation declining rapidly (or becoming impossible) once trees attain mature status. Such a system just described, involving slow-growth storage of shoots of juvenile clones in conjunction with concurrent evaluation and selection from replicated field-grown clones at a stage when desirable adult characteristics become apparent, may therefore minimize the difficulties usually encountered during propagation of mature trees.

8.3.6 *In vitro* methods for collecting germplasm in the field

Recently, much interest has been shown in the development of specialized *in vitro* techniques to facilitate collection of plant germplasm in the field. Placing tissue explants taken from tree specimens in the field directly into tissue culture has several distinct advantages over conventional sampling methods (Withers, 1987). These may be briefly summarized as follows:

1. Collected tissue samples can be placed into *in vitro* culture within minutes of collection, so facilitating optimal survival of materials prone to rapid deterioration.
2. Material may be collected from trees of species where only vegetative explants (shoots/buds, suckers, storage organs, etc.) are normally available.
3. Vegetative material could be collected when exact clonal genotypes are required.
4. Collecting mature embryos of recalcitrant seeds would achieve the same aim (i.e. of germplasm conservation), and at the same time reduce bulk. Such a procedure would have obvious applications for species such as coconut (*Cocos nucifera*).

5. Collecting immature seeds or immature embryos of both orthodox and recalcitrant seeds would enable such materials to be gathered and supported physiologically despite their being delicate.
6. Techniques of *in vitro* collection in the field may facilitate or make feasible the collection of gynaecia for later *in vitro* fertilization, and also pollen and/or anthers at the optimal stage of development for subsequent haploid production *in vitro*.

The collection of embryos of recalcitrant-seed species should facilitate the conservation *in vitro* of tree crops whose seed is extremely short-lived, and which loses viability within only a few days, as in the case of mahogany (Vivekanandan, 1978) and rambutan (*Nephelium lappaceum*) (Ellis, 1984).

Collection of immature seeds/embryos will enable sampling of tree specimens to be carried out even though the collecting expedition may not necessarily reach the site at the time when seed is at optimum maturity. In addition, when collecting trips are undertaken to distant locations or inaccessible sites, the collection of actively growing embryos or shoot-tip material will enable departure from the site to be delayed until a sufficient number of storage propagules are already developing satisfactorily *in vitro*.

Most *in vitro* methods of germplasm collection in the field that have been proposed to date involve mostly a refinement of standard methods of *in vitro* culture initiation and subsequent propagation. However, *in vitro* collecting techniques usually require additional measures to be taken in order to control residual microbial contamination. This normally involves inclusion of antibiotic and/or fungicide in the initiation medium. Levels of contamination encountered through the application of *in vitro* collecting techniques are of the order of 6–10%. A recent report by Withers (1987) has discussed in detail the methodology involved in the *in vitro* collection of plant germplasm resources in the field.

In vitro collection techniques have been successfully employed for collection of coconut palm (*Cocos nucifera*) embryos in the field (Assy Bah *et al.*, 1987), in addition to embryos of other palm species such as *Caryota urens*, *Phoenix sylvestris*, *Jubea chilensis* and *Butia capitata* (Sossou *et al.*, 1987). Yidana *et al.* (1987) have recently applied similar techniques for collection of shoot nodal cuttings of cocoa (*Theobroma cacao*). However, it was pointed out in the latter report that the lack of a satisfactory *in vitro* propagation technique for cocoa means that additional development of *in vitro* techniques to enable processing of shoots after collection is required, such as the use of *in vitro* micrografting techniques. Sossou *et al.* (1987) have stated that *in vitro* explanting in the field of coconut palm embryos will enable selection of nuts with elite kernels (the primary economic product of the plant) during collection, with subsequent excision and collection of embryos only from those selected nuts which show desirable kernel qualities.

8.3.7 Pollen storage

Various workers have demonstrated the feasibility of storing pollen of woody species at both low ($-20°C$) and ultralow ($-196°C$) temperatures. Techniques of pollen storage have been applied to several woody crops, including *Diospyros kaki* (Wakisaka, 1964), *Citrus grandis*, *C. hassaku* and *C. natsudaidai* (Kobayashi *et al.*, 1978), *C. limon* (Ganeshan and Sulladmath, 1983), various *Prunus* spp. (Parfitt and Almehdi, 1984) and *Carica papaya* (Ganeshan, 1986). However, such techniques, although of possible value in plant-breeding programmes, do not conserve the whole genotype of a plant and therefore have only a complimentary role in the long-term conservation strategy of woody species (Stushnoff and Fear, 1985).

8.4 STABILITY OF TREE TISSUE-CULTURE SYSTEMS

It is an obvious requirement of any tissue-culture system potentially employed for purposes of plant germplasm preservation that such a system be genetically stable. The genetic integrity of the culture must be guaranteed during the routine procedures of repeated (over months or years) propagule multiplication. In a review of chromosome-number variation in plants regenerated from tissue culture, D'Amato (1978) concluded that: 'chromosome number variation occurs in practically all calli and suspension cultures'. The author further stated: 'the only means of ensuring genetic stability, essential for clonal propagation, is plant regeneration from shoot meristem cultures'. Unfortunately, as discussed previously, whereas for most important tree crop species clonal multiplication is feasible by the 'multiple-shoot technique', most *in vitro* propagation processes developed to date for palms have involved a callus stage.

Early workers engaged in the development and subsequent application of *in vitro* methods to palm propagation were of the opinion that such systems were reasonably stable, even though a callus phase was involved. It was assumed that, in terms of plant regeneration from a disorganized culture system, tissues of species such as oil palm were at the more stable end of the spectrum. In a review of palm tissue culture, Choo *et al.* (1982) reported the identification of only four abnormal palm plants out of a total of 13 000 trees produced by *in vitro* propagation, and subsequently established in field trials/nurseries. Similarly, Ulrich *et al.* (1982), utilizing techniques of isoenzyme electrophoresis, found no evidence of variation in palm plants regenerated from both cryopreserved and non-frozen embryogenic callus cultures of date palm.

Recently, however, an article by Corley *et al.* (1986) has reported that, whereas the first palms produced by *in vitro* culture showed no abnormal flower development, those palm plants regenerated from the same callus-culture lines maintained *in vitro* for three years showed abnormal flower development (a presumed indicator of genetic aberration) in 88–95% of cases, depending on the clone. Engelmann *et al.* (1987) have proposed that these occurrences appear to be associated with the conditions prevalent during the *in vitro* phase of the propagation procedure, especially when exogenous growth regulators are used. The latter authors also point out that the embryoid clones produced by ORSTOM/IRHO in France have been maintained and subcultured on a medium devoid of growth hormones for several years, and all palms produced using the ORSTOM/IRHO technique show normal development.

A recent review by Jarret (1986) dealing with *in vitro* propagation techniques applied to *Musa* genotypes has also proposed the possibility that certain conditions/substances encountered during *in vitro* culture, e.g. the presence and concentration of media constituents such as the cytokinin compound benzylaminopurine (BAP), may influence the stability of sub-sequent regenerants. A relatively high rate of variation has been observed in *Musa* clones propagated via shoot-tip cultures. This has been discussed by Reuveni *et al.* (1985). The mutation rate appears to vary between 4% and 14% for phenotypic characters observed in the field and consi-dered to be genotypically controlled (Jarret, 1986). The most frequently observed variants are those which affect plant height and pigmentation. *In vitro* propagated plants of *Musa* clones tend to show various juvenile characteristics (Hwang *et al.*, 1984). Also, in at least one instance, variation in plant pigmentation and fruit-bunch characteristics was the result of the unmasking of chimerism in the parent plant. Jarret (1986) suggests that shoot production from *in vitro* cultures of *Musa* genotypes may occur by either of two pathways, and that regulation of these pathways may affect the genetic integrity of the resultant plantlets. The first pathway involves the development of shoots from axillary buds, which are usually consid-ered to be genetically stable. However, another type of budding leading to (potentially adventitious) shoot development has been described by vari-ous workers (e.g. Vulysteke and de Langhe, 1985; Banerjee *et al.*, 1986). In this pathway, bulbil-like structures initially develop, which in some cul-tures continue to develop into a whitish tissue, covered on its surface with numerous minute meristems. These minute meristems have been observed by numerous investigators and have been suggested to be ad-ventitious in origin. In addition, their occurrence is apparently linked to the propagation rate of the culture system in which they occur, which in turn depends on the concentration of cytokinin in the culture medium.

Banerjee *et al.* (1986) have described in detail the process of shoot formation in *Musa* shoot-tip cultures. With most cultivars studied, the presence of a 1 μM concentration of BAP resulted only in shoot development with no proliferation, and rooted plantlets were regenerated and established *in vitro*. However, a ten-fold increase in the concentration of cytokinin (i.e. BAP) in the medium resulted in the *de novo* formation of multiple meristems at the base of the central leaf primordium. It was noted that the rate of multiple bud proliferation was dependent on the particular cultivar under study, and proposed that the rate may be linked to the genomic constitution of the cultivar, since possession of one or two B genomes appeared to correlate with greater bud proliferation.

Jarret (1986) points out that it is not possible to speculate whether genetic variation among plants derived from these apparently adventitious buds will be greater than among plants derived from non-adventitious axillary buds or meristems, although several field studies are now in progress (Cronauer and Krikorian, 1985; Reuveni *et al.*, 1985).

Jarret (1986) concludes that any tissue-culture system of propagation involving adventitious bud regeneration should be avoided, since variation arising through *de novo* bud development may give rise to mutants through somaclonal variation. In addition, such an *in vitro* system could obviously not be recommended as the basis for a method of tissue-culture germplasm storage, since the genetic stability of genotypes during storage could not be guaranteed.

8.4.1 Rejuvenation of woody tissues *in vitro*

Much attention has been given in recent years to the phenomenon of rejuvenation of mature woody-plant tissues during prolonged periods of *in vitro* culture. Various workers have noted that *in vitro* shoot cultures of woody species become progressively easier to propagate with increasing time in culture, and have attributed such an occurrence to a reversion to the juvenile phase.

The fact that the potential ease of propagation of a woody species is largely dependent on the physiological status (i.e. whether juvenile or mature) of the material in question has been discussed in detail earlier in this chapter. During the initiation phase of shoot-culture establishment, various workers have found that the propensity of a particular meristem or shoot apex to form a culture is dependent on the age and physiological status of the parent tree from which the explant is derived, in addition to the size of the explant used. The region of tree from which explant material originates may also influence the ease of culture establishment, since some particular zones of a tree maintain a physiologically juvenile status throughout the life of the plant (Westwood, 1978; Bonga, 1982b).

Jones (1978) compared the case of culture initiation of temperate fruit species from both one-year-old and eight-year-old trees. He found that, whereas shoot apices from one-year-old (i.e. physiologically juvenile) trees were easy to establish, for eight-year-old (mature) trees there existed a 'critical explant size' of 3 mm, and it was impossible to initiate cultures from shoot apices of less than this size. Also, there was often a long delay before shoot tips from the older trees would commence growth. Vertesy *et al.* (1980) reported that it was impossible to initiate cultures using true meristem explants taken from greenhouse-grown plants of the plum rootstock Pixy (*Prunus insititia*). However, once proliferating shoot cultures had been established from large shoot tips, then passed through several subculture cycles, it proved relatively easy to grow true meristems (0.1–0.2 mm) dissected from the *in vitro* cultured shoots. More recently, Monteuwis (1986) was able to graft meristems dissected from 100-year-old trees of *Sequoiadendron giganteum* successfully onto *in vitro* grown seedlings, and thereby initiate shoot growth in culture. It was noted that shoot growth from grafted apices displayed juvenile characteristics. Such a technique of *in vitro* micrografting may prove of value in the future as a means of culture establishment for elite specimens of mature woody species. In addition to noting an increased ease of shoot proliferation with increasing time in culture (Hutchinson, 1982), many workers have provided evidence that the capacity of *in vitro* shoots to initiate adventitious roots also increases with increasing time in culture. For example, Sriskandarajah *et al.* (1982) found that, with the apple scion cultivar Delicious, there was a progressive improvement in the rooting of the shoots with increasing numbers of subcultures. After four subcultures, the percentage rooting was 21, whilst after 31 subcultures the percentage rooting had risen to 79.

This phenomenon of enhanced ease of rooting with increased degree of juvenility parallels the *in vivo* situation discussed earlier. Further evidence that the process of prolonged *in vitro* culture leads to rejuvenation has been provided by several workers, who noted the occurrence of shoots/plants with juvenile characters in shoot cultures of various woody crop species such as grape (Mullins *et al.*, 1979), blueberry (Lyrene, 1981) and *Musa* (Hwang *et al.*, 1984).

A recent report by Barghchi (1987a) has described the occurrence of rejuvenation in shoot cultures of the subtropical fruit crop tamarillo or 'tree tomato'. Cultures initiated using explants from mature plants eventually reverted to the juvenile state, and the time of onset of rejuvenation was found to be associated with the frequency of regeneration, i.e. a more frequent subculture cycle induced the early production of shoots with juvenile characteristics. It was also noted that shoots produced adventitiously were rejuvenated in the first generation.

David (1982) has proposed that *in vitro* rejuvenation could aid in the propagation of mature Gymnosperms. Cuttings of *Sequoia sempervirens* rooted by conventional means were compared with shoots rooted *in vitro* after passing through four subculture cycles with exposure to activated charcoal. Conventional cuttings were difficult to root and displayed pronounced plagiotropic growth, whereas *in vitro* propagated shoots showed an enhanced capacity for rapid rooting; and subsequent growth after *in vivo* establishment was orthotropic, indicating that a rejuvenation of the material had occurred *in vitro*.

Recently, the occurrence of phase reversion during *in vitro* propagation has given cause for concern, since various workers have demonstrated that bud production in proliferating shoot cultures of woody species may be (at least partially) by an adventitious pathway, and not from the outgrowth of preformed axillary meristems, as previously assumed.

A report by Welander (1985) has discussed the occurrence of two types of adventitious structures in shoot cultures of the apple cultivar Akero, both of which were capable of giving rise to shoots. One of these structures consisted of meristematic areas forming shoot primordia which eventually developed into adventitious shoots. These shoot primordia were shown to arise from within a region of firm callus that surrounded the base of the explanted shoot stem. The other type of structure was described by the author as having a considerable superficial resemblance to specialized woody structures known as sphaeroblasts.

Various workers have described the *in vivo* occurrence of sphaeroblasts, and the fact that shoots arising from such structures tend to display juvenile characteristics (Wellensiek, 1952; Hatcher and Garner, 1955). Bonga (1987) has also recently discussed the fact that, with most woody species, shoots arising by an adventitious (or *de novo*) pathway, whether obtained *in vivo* (from sphaeroblasts) or *in vitro*, are physiologically juvenile. It follows that the occurrence of shoots arising *de novo* from proliferating shoot-tip cultures of woody species may thus explain the phenomenon of phase reversion during repeated subculturing. Although the factors which govern the initial production of adventitious shoots in organized proliferating shoot cultures are not fully known, the presence of, or continuous exposure of shoots to, high levels of cytokinins such as BAP, may promote such an occurrence. This proposal has already been discussed earlier in this section, in relation to *Musa* spp. and oil palm (*Elaeis guineensis*).

The occurrence of adventitious shoot production has also been observed in proliferating shoot cultures of the apple scion cultivar Greensleeves, which had been repeatedly subcultured in the presence of BAP (Wilkins and Dodds, unpublished observations). When single shoots were explanted to fresh medium containing BAP, swellings around the basal region of the stem became apparent after several days of culture. Eventually, after elongating during 3–4 weeks of further growth, the outgrowths developed foliar appendages and thereafter assumed normal shoot appearance. Longitudinal sections of the base of the stem taken after 14 days of culture clearly show *de novo* differentiation of 'organized centres' within the stem cortex (Figure 8.5). The production of shoots by an adventitious pathway also raises the question of whether such shoots are genetically stable, and further research is needed in this area to determine whether a relationship exists between the number of subculture passages and exogenous cytokinin supply, and the subsequent occurrence of *de novo* shoot production. Research effort is needed to ascertain whether woody plants propagated *in vitro* from cultures continuously producing shoots over long periods of time are genetically stable. There has been little work conducted to date regarding whether high concentrations of other cytokinin compounds such as kinetin or 2-isopentyl-adenine (2iP) may also promote adventitious shoot production.

8.4.2 Assessment of genetic stability of woody tissues *in vitro*

Applicable methods of assessing the stability of plant tissue cultures utilized for purposes of germplasm storage are reviewed in Chapter 4. Due to the limited number of published reports that have described the successful germplasm storage of woody species, the scope for application of such techniques to date has obviously been limited. One example is the report of Ulrich *et al.* (1982), which used isoenzyme markers to assess the stability of cryopreserved date palm callus. The use of isoenzymes as genetic markers for variety identification and characterization of germplasm has received considerable attention (Tanksley and Orton, 1983). Simpson and Withers (1986) have recently compiled a comprehensive literature review dealing with characterization of plant material using isooenzyme electrophoresis. Several workers have recently applied techniques of isoenzyme electrophoresis for the characterization of germplasm of woody crop species such as cocao (*Theobroma cacao*) (Atkinson *et al.*, 1986). Isoenzymes have also been used as genetic markers in banana and plantains (*Musa* spp.) (Jarret and Litz, 1986), and to identify plum × peach hybrids (Parfitt *et al.*, 1985). Such reports indicate that techniques of isoenzyme analysis are already available for assessing the stability of woody plants stored *in vitro*. With other crop species, similar techniques have already been applied to identify mutants arising during *in vitro* culture. For example, Brettell *et al.* (1986) used isoenzymes to detect a point mutation in the alcohol dehydrogenase gene in regenerated maize plants. However, the potential use of isoenzyme analysis to detect mutants is only feasible if the point at which the particular mutation occurs is within the particular gene sequence that codes for an enzyme that can be assayed for. Withers (1988) has very

Figure 8.5 Longitudinal section of basal region of a single shoot of the EMLA apple scion cultivar Greensleeves (*Malus domestica*). Shoots had been cultured for 14 days on agar-solidified medium containing benzylaminopurine, at 25°C. The section shows the presence of 'organized centres' (arrowed), developing *de novo* within the cortex of the swollen regions of the stem base. Subsequently, swellings elongate and eventually develop foliar appendages, thereafter assuming normal shoot appearance.

recently pointed out a possible drawback concerning the use of isoenzyme screening for assessing genetic stability, since authors such as Kevers and Gasper (1985) have found a correlation between the physiological phenomenon of 'vitrification' in *in vitro* shoot cultures, and changes in peroxidase isoenzyme profiles. Recently, methods of recombinant DNA technology, such as analysis of restriction fragment length polymorphisms (RFLPs) in conjunction with an appropriate probe, have been proposed as a means of detecting mutations arising through rearrangements to the genome, such as small translocations, deletions, etc.

Although several studies are presently in progress to investigate the efficacy of RFLPs as a means of monitoring the *in vitro* stability of cultures of various herbaceous species such as *Solanum tuberosum* (potato), to this author's knowledge these techniques have not yet been applied to *in vitro* cultures of woody crop species.

Bernatzky and Tanksley (1989) have very recently discussed the use of restriction fragments as molecular markers for germplasm evaluation and utilization, and given brief details of the methodologies employed.

Another facet of the requirement for assessing stability of woody tissues *in vitro* is the question of whether any of the aforementioned techniques would be able to detect subtle changes in the physiological status of woody plant cultures. It would be of value to develop a method of assessing the degree of juvenility of *in vitro* cultures of woody species, in relation to the time in culture (i.e. number of prior subculture cycles) and the type and concentration of growth regulators within the medium. Such methods ought to detect changes that may precede the onset of *de novo* shoot formation, and the concurrent risk of subsequent genetic aberrations.

8.5 CONCLUSION

In conclusion, it is evident that great progress has already been made concerning the application of tissue-culture techniques to the propagation and also germplasm storage of woody species. For some tree crops, tissue-culture propagation is now the norm, and new reports are constantly appearing in the literature describing the successful propagation *in vitro* of woody crop species. The potential for establishment of *in vitro* gene banks for certain species of woody crops already exists. In addition, the large amount of as yet unpublished information currently contained in computer databases, and relating to the large number of woody species presently under investigation with a view to developing *in vitro* germ-plasm-preservation methods, indicates that such techniques will become available for many other woody crops in the near future.

However, in spite of the numerous successful applications of tree tissue-culture methods, many problems still exist, especially concerning the establishment of cultures and subsequent plant regeneration from mature Gymnosperm species. Also, for species such as *Citrus* spp., avocado (*Persea americana*) and cocoa (*Theobroma cacao*), it has proved difficult to develop efficient systems of *in vitro* propagation, although the development of refined tissue-culture techniques such as *in vitro* shoot-tip grafting may eventually overcome many of these problems. The very recent development and subsequent demonstration of the effectiveness of *in vitro* techniques for collecting germplasm in the field should overcome many previously encountered problems relating to the collection of delicate or bulky specimens in remote and inaccessible regions, so ensuring the continued existence of endangered tree resources for use by future generations.

ACKNOWLEDGEMENTS

I wish to thank Professor E. G. Cutter for reviewing the manuscript of this chapter, and for her valuable and instructive comments.

In addition to those workers (not named below) who supplied me with preprints or reprints of their work, I wish to express my sincere thanks to the following scientists who kindly replied to my enquiries and provided unpublished information: Dr M. Barghchi, Dr F. Engelmann, Dr R. Jarret, Dr T. R. Marks, Dr A. F. Mascarenhas, Dr L. Navarro, Dr A. R. Rosten, Dr J. F. Seelye, Prof. C. Stushnoff and Dr A. M. Vietez.

I would like to thank Dr L. A. Withers for conducting a search of the IBPGR *In Vitro* Conservation Database.

I thank my wife Blanca for her help and patience during the preparation of this chapter.

Miss J. Cox, Mr R. Gregory and Mr I. Miller are thanked for photographic assistance.

REFERENCES

Abbott, A. J. and Whiteley, E. (1976) Culture of *Malus* tissues *in vitro*. I. Multiplication of apple plants from isolated shoot apices. *Sci. Hort.*, **4**, 183–189.

Abdul Rahman, N., Diner, A. M., Karnosky, D. F. and Skilling, D. D. (1986) in *Proceedings VI International Congress of Plant Tissue and Cell Culture*, Vol. 1 (ed. D. A. Somerers, B. G. Genenback, D. D. Biesboer, W. P. Hackett and C. E. Green), International Association for Plant Tissue Culture, Minneapolis, p. 401.

Ahee, J., Arthuis, P., Cas, G., Duval, Y., Guenin, G., Hanower, J., Hanower,

P., Lievoux, D., Lioret, C., Malaurie, B., Pannetier, C., Raillot, D., Varechon, C. and Zuckerman, L. (1981) La multiplication vegetative *in vitro* du palmier a huile par embryogenese somatique. *Oleagineaux*, **36**(3), 113–118.

Aitken-Christie, J. and Singh, A. P. (1987) in Bonga, J. M. and Durzan, D. J. (eds.). *Cell and Tissue Culture in Forestry*, Vol. 2, *Specific Principles and Methods: Growth and Developments*, Martinus Nijhoff Publishers, pp. 285 –304.

Amin, M. N. and Jaiswal, V. S. (1987) Rapid clonal propagation of guava through *in vitro* shoot proliferation in nodal explants of mature trees. *Plant Cell Tissue and Organ Culture*, **9**(3), 235–243.

Anderson, W. C. (1975) Propagation of rhododendrons by tissue culture: Part 1. Development of a culture medium for multiplication of shoots. *Comb. Proc. Int. Plant Prop. Soc.*, **25**, 129–135.

Anderson, W. C. (1978a) Tissue culture propagation by rhododendrons. *In vitro*, **14**, 334 (abstract).

Anderson, W. C. (1978b) Rooting of tissue cultured rhododendrons. *Comb. Proc. Int. Plant. Prop. Soc.*, **28**, 135–139.

Anderson, W. C. (1980) in *Proc. Conf. Nursery Production of Fruit Plants through Tissue Culture – Applications and Feasibility*, US Dept Agri SEA. ARR-NE-11, pp. 27–34.

Anonymous (1982) *Fruit-bearing Forest Trees*, FAO Forestry Paper No. 34, Food and Agriculture Organization of the United Nations (FAO), Rome.

Anonymous (1983) *Food and Fruit-bearing Forest Species 1: Examples from Eastern Africa*, FAO Forestry Paper No. 44/1, Food and Agriculture Organization of the United Nations (FAO), Rome.

Anonymous (1984) *Food and Fruit-bearing Forest Species 2: Examples from Southeastern Asia*, FAO Forestry Paper No. 44/2, Food and Agriculture Organization of the United Nations (FAO), Rome.

Anonymous (1986) *Food and Fruit-bearing Forest Species 3: Examples from Latin America*, FAO Forestry Paper No. 44/3, Food and Agriculture Organization of the United Nations (FAO), Rome.

Assy Bah, B., Durand-Gasselin, T., Pennetier, C. and Buffard-Morel, J. (1987) A simple *in vitro* method for collecting coconut embryos in the field. *FAO/IBPGR Pl. Genetic Resources Newsletter*, **70** (in preparation).

Atkinson, M. D., Withers, L. A. and Simpson, M. J. A. (1986) Characterisation of cacao germplasm using isoenzyme markers. I. A preliminary survey of diversity using starch gel electrophoresis and standardization of the procedure. *Euphytica*, **35**, 741–750.

Attree, S. M., Bekkaoui, F., Dunstan, D. I. and Fowke, L. C. (1987) Regeneration of somatic embryos from protoplasts isolated from an embryogenic suspension culture of white spruce (*Picea glauca*). *Plant Cell Rpts*, **6**, 480–483.

References

Attree, S. M., Dunstan, D. I. and Fowke, L. C. (1989) Plantlet regeneration from embryogenic protoplasts of white spruce (*Picea glauca*). *Bio/Technology* (in press).

Bajaj, Y. P. S. (1984) Induction of growth in frozen embryos of coconut and ovules of *Citrus*. *Current Sci.*, **53**(22), 1215–1216.

Bajaj, Y. P. S. (1986) *Biotechnology of Agriculture and Forestry, Trees: 1*, Springer Verlag, Berlin.

Ball, E. A. (1950) Differentiation in a callus culture of *Sequoia sempervirens*. *Growth*, **14**, 295–325.

Banerjee, N. and de Langhe, E. (1985) A tissue culture method for rapid clonal propagation and storage under minimal growth conditions of *Musa* (banana and plantain). *Plant Cell Rpts*, **4**, 351–354.

Banerjee, N., Vuylsteke, D. and de Langhe, E. A. L. (1986) in *Plant Tissue Culture and its Agricultural Applications* (ed. L. A. Withers and P. G. Alderson), Butterworths, London, pp. 139–148.

Bapat, V. A. and Rao, P. S. (1988) Sandalwood plantlets from 'synthetic seeds'. *Plant Cell Rpts*, **7**, 434–436.

Barghchi, M. (1987a) in *Book of Abstracts, 2nd Annual Conference of the International Plant Biotechnology Network (IPBNet) and Tissue Culture for Crops Project (TCCP)*, Thailand, 1987.

Barghchi, M. (1987b) in *Book of Abstracts, IAPTC – 7th Biennial Conference of the New Zealand IAPTC Branch*, Palmerston North, New Zealand.

Barghchi, M. (1987c) in *International Association for Plant Tissue Culture (IAPTC) – 7th Biennial Conference of New Zealand Branch, Book of Abstracts*.

Barlass, M. and Skene, K.G.M. (1983) Long-term storage of grape *in vitro*. *Int. Board Plant Genet. Res. Newletter*, **53**, 19–21.

Bayliss, M. W. (1980) Chromosomal variation in plant tissues in culture. *Int. Rev. Cytol. Suppl.*, **11A**, 113–144.

Bekkaoui, F., Saxena, P. K., Attree, S. M., Fowke, L. C. and Dunstan, D. I. (1987) The isolation and culture of protoplasts from an embryogenic cell suspension culture of *Picea glauca* (Moench) Voss. *Plant Cell Rpts*, **6**, 476–479.

Bekkaoui, F., Pilon, M., Laurie, E., Raju, D. S. S., Crosby, W. L. and Dunstan, D. I. (1988) Transient gene expression in electroporated *Picea glauca* protoplasts. *Plant Cell Rpts*, **7**, 471–484.

Bennet, E. (1968) *FAO/IBP Technical Conference on the Exploration, Utilisation and Conservation of Plant Genetic Resources*, FAO, Rome.

Berg, L. A. and Bustamente, M. (1974) Heat treatment and meristem culture for the production of virus-free bananas. *Phytopathology*, **64**, 320–322.

Bernatzky, R. and Tanksley, S. D. (1989) in *The Use of Plant Genetic Resources* (ed. A. H. D. Brown, O. H. Frankel, D. R. Marshall and J. T. Williams, Cambridge University Press, Cambridge, pp. 353–376.

Bettencourt, E. J. and Konopka, J. (1989) *Directory of Germplasm Collections. 6.II. Industrial Crops: Actinidia, Amelanchier, Canja, Castanea, Corylus, Cydonia, Diospyros, Fragaria, Juglans, Malus, Mespilus, Morus, Olea, Pistacia, Punica, Prunus, Pyrus, Ribes, Rosa, Rubus, Sambucus, Sorbus, Vaccinium and Others*, International Board for Plant Genetic Resources, Rome.

Bhojwani, S. S., Dhawan, V. and Cocking, E. C. (1986) *Plant Tissue Culture – a Classified Bibliography*, Oxford, Elsevier.

Biondi, S. and Thorpe, T. A. (1982) in *Tissue Culture of Economically Important Plants*, (ed. A. N. Rao), COSTED, Singapore, pp. 197–204.

Blake, J. (1983) in *Tissue Culture of Trees*, (ed. J. H. Dodds), Croom-Helm, London, pp. 29–50.

Blake, J. and Eeuwens, C. J. (1982) in *Tissue Culture of Economically Important Plants*, (ed. A. N. Rao), COSTED, Singapore, pp. 145–148.

Blake, R. O. (1984) Moist forests of the tropics. A plea for protection and development. *World Res. Inst. J.*, 34–39.

Bonga, J. M. (1982a) in *Tissue Culture of Economically Important Plants* (ed. A. N. Rao), COSTED, Singapore, pp. 191–196.

Bonga, J. M. (1982b) in *Tissue Culture in Forestry*, Martinus Nijhoff, London, pp. 387–412.

Bonga, J. M. (1987) in *Advances in Cell Culture*, Vol. 5, Academic Press, New York, pp. 209–239.

Bonga, J. M. and Durzan, D. J. (1987). *Cell and Tissue Culture in Forestry*, vol. 3, *Case Histories: Gymnosperms, Angiosperms and Palms*, Martinus Nijhoff Publishers.

Borchert, R. (1976) The concept of juvenility in woody plants. *Acta Hort.*, **56**, 21.

Branton, R. L. and Blake, J. (1983) Development of organised structures in callus derived from explants of *Cocos nucifera* L. *Ann. Bot.*, **52**, 673–678.

Brettell, R. I. S., Dennis, E. S., Scowcroft, W. R. and Peacock, W. J. (1986) Molecular analysis of a somaclonal mutant of maize alcohol dehydrogenase. *Mol. Gen. Genet.*, **202**, 235–239.

Brookes, H. J. and Barton, D. W. (1977) A plan for national fruit and nut germplasm repositories. *Hort. Sci.*, **12**, 298–300.

Broome, O. C. and Zimmerman, R. H. (1978) *In vitro* propagation of blackberry. *Hort. Sci.*, **13**, 151–153.

Brown, A. G. (1975) in *Advances in Fruit Breeding* (ed. J. Janick and J. N. Moore), Purdue University Press, West Lafayette, Indiana, pp. 3–37.

Brown, C. L. and Sommer, H. E. (1982) in *Tissue Culture in Forestry* (ed. J. M. Bonga and D. J. Durzan), Martinus Nijhoff, The Hague, pp. 109–142.

Brown, S. and Lugo, A. E. (1985) in *Environmental Science: A Framework for Decision Making*, (ed. D. D. Chiras), Benjamin Cummings, Menlo Park, California.

References

Burley, J. and Namkoong, G. (1980) *Conservation of Forest Genetic Resources*, XI Commonwealth Forestry Conference, Trinidad.

Cannell, M. G. R. and Last, F. T. (1976) *Tree Physiology and Yield Improvement*, Academic Press, New York.

Carlisle, A. and Teich, A. H. (1971) The costs and benefits of tree improvement programs. *Can. For. Serv. Publ.*, **1302**, 14.

Carlos, J. T. Jr (1979) *Conservation Genetic Resources of Coconut*, PCRDF Professional Lect. Aug. 21, 1979, UPLB, Laguna, Philippines.

Caswell, K. L., Taylor, N. J. and Stushnoff, C. (1986) Cold hardening of *in vitro* apple and Saskatoon shoot cultures. *Hort. Sci.*, **21**(5), 1207–1209.

Cauvin, B. and Marien, J. N. (1978) La multiplication vegetative des *Eucalyptus* en France. *Ann. Rech. Sylvicoles*, 141–175.

Chaturvedi, H. C. and Mitra, G. C. (1974) Clonal propagation of Citrus from somatic callus cultures. *Hort. Sci.*, **9**, 118–220.

Cheema, G. S. (1989) Somatic embryogenesis and plant regeneration from cell suspensions and tissue cultures of mature Himalayan poplar (*Populus ciliata*). *Plant Cell Rpts*, **8**, 124–127.

Cheng, T. Y. (1980) in *Proc. Conf. Nursery Production of Fruit Plants through Tissue Culture–Applications and Feasibility*, US Dept. Agri. SEA. APR-NE-11.

Chippendale, G. M. (1976) Eucalyptus nomenclature. *Aust. For. Res.*, **7**, 69–107.

Choke, H. C. (1973) in *Survey of Crop Genetic Resources in their Centres of Diversity* (ed. O. H. Frankel), FAO/IBP, Rome, pp. 160–164.

Choo, W. K., Yew, W. C. and Corley R. H. V. (1982) in *Tissue Culture of Economically Important Plants* (ed. A. N. Rao), COSTED, Singapore, pp. 138–144.

Copes, D. L. (1976) Comparative leader growth of Douglas-fir grafts, cuttings, and seedlings. *Tree Planters Notes*, **27**, 13.

Corley, R. H. V., Barrett, J. N. and Jones, L. H. (1977) in *International Developments in Oil Palm* (ed. D. A. Earp and W. Newall), Incorporated Society of Planters, Kualur Lumpur.

Corley, R. H. V., Wooi, K. C. and Wong, C. Y. (1979) Progress with vegetative propagation of oil palm. *Planter*, **55**(641), 377–380.

Corley, R. H. V., Lee, C. H., Law, I. H. and Wong, C. I. (1986) Abnormal flower development in oil palm clones. *Planter*, **62**, 233–240.

Crassweller, R. M., Feree, D. C. and Nicholas, L. P. (1980) Flowering crab apple as potential pollinisers for commercial apple cultivars. *J. Am. Soc. Hort. Sci.*, **105**, 453–457.

Cremer, K. W. (1969) Growth of Eucalyptus in experimental plantations near Canberra. *Aust. For.*, **33**, 2.

Cronauer, S. S. and Krikorian, A. D. (1984) Multiplication of *Musa* from excised stem tips. *Ann. Bot.*, **53**, 321–328.

Cronauer, S. S. and Krikorian, A. D. (1985) in *Biotechnology in Agriculture and Forestry. I. Trees* (ed. Y. P. S. Bajaj), Springer-Verlag, Berlin, pp. 233 −252.

Cross, M. (1988) Spare the tree and spoil the forest. *New Scientist*, **121**, 24−25.

D'Amato, F. (1975) in *Crop Genetic Resources for Today and Tomorrow* (ed. O. H. Frankel and J. G. Hawkes), Cambridge University Press, Cambridge, pp. 333−348.

D'Amato, F. (1978) in *Frontiers of Plant Tissue Culture, 1978* (ed. T. A. Thorpe), IAPTC, Calgary, pp. 287−296.

Datta, S. K., Datta, K. and Pramanik, T. (1982) *In vitro* clonal multiplication of mature trees of *Dalbergia sissoo* Roxb. *Plant Cell Tissue and Organ Culture*, **2**, 15−20.

David, A. (1982) in *Tissue Culture in Forestry* (ed. J. M. Bonga and D. J. Durzan), Martinus Nijhoff Publishers, The Hague, pp. 72−108.

Davidson, J. (1977) *Problems of Vegetative Propagation of Eucalyptus*, FAO Third World Consultation on Forest Tree Breeding, Canberra, Australia, pp. 857−882.

De Candolle, A. (1889) *Origin of Cultivated Plants*, Kegan Paul Trench and Co., London.

de Langhe, E. A. L. (1984) in *Crop Genetic Resources: Conservation and Evaluation* (ed. J. H. W. Holden and J. T. Williams), London, George Allen and Unwin, pp. 131−137.

Delbard, H. (1982) Micropropagation of roses at Delbard Nurseries. *The Plant Propagator*, **28**(3), 7−8.

Doorenbos, J. (1965) in *Encyclopedia of Plant Physiology* (ed. W. Ruhland), Vol. 15 (Pt 1), Berlin, Springer-Verlag.

Dore-Swamy, R., Srinivasa Rao, N. K. and Chacko, E. K. (1983) Tissue culture propagation of banana. *Sci. Hort.*, **18**, 247−252.

Doughty, S. and Power, J. B. (1988) Callus formation from leaf mesophyll protoplasts of *Malus domestica* Borkh. cv. Greensleeves. *Plant Cell Rpts*, **7**, 200−201.

Driver, J. A. and Kuniyuki, A. H. (1984) *In vitro* propagation of Paradox walnut rootstock. *Hort. Sci.*, **19**, 507−509.

Duhoux, E., Sougoufara, B. and Dommergues, Y. (1986) Propagation of *Casuarina equisetifolia* through axillary buds of immature female inflorescences cultured *in vitro*. *Plant Cell Rpts*, **3**, 161−164.

Durand-Cresswell, R., Boulay, M. and Franclet, A. (1982) in Bonga, J. M. and Durzan, D. J. (eds.). *Tissue Culture in Forestry* (ed. J. M. Bonga and D. J. Durzan), Martinus Nijhoff, London, pp. 150−181.

Durzan, D. J. (1982) in (ed. J. M. Bonga and D. J. Durzan), *Tissue Culture in Forestry* (ed. J. M. Bonga and D. J. Durzan), Martinus Nijhoff, The Hague, pp. 36−71.

Durzan, D. J. (1985) in *Tissue Culture in Forestry and Agriculture* (ed. R. R. Henke, K. W. Hughes, M. J. Constantin and A. Hollaender), Plenum Press, New York, pp. 233–256.

Dutcher, R. D. and Powell, L. E. (1972) Culture of apple shoots from buds in vitro *J. Am. Soc. Hort. Sci.*, **97**, 511–514.

Eckholm, E. (1975) *The Other Energy Crisis: Firewood*, Worldwatch Paper 1, Worldwatch Institute, Washington DC.

Elliot, R. F. (1972) Axenic culture of shoot apices of apple. *N. Z. J. Bot.*, **10**, 254–258.

Ellis, D., Roberts, D., Sutton, B., Lazaroff, W., Webb, D. and Flinn, B. (1989) Transformation of white spruce and other conifer species by *Agrobacterium tumefaciens*. *Plant Cell Rpts*, **8**, 16–20.

Ellis, R. H. (1984) Revised table of seed storage characteristics. *FAO/IBPGR Plant Gen. Resources Newsletter*, **58**, 16–33.

Engelmann, F. (1985) Biotechnologies (Plant Physiology) – Survival and proliferation of oil palm (*Elaeis guineensis* Jacq.) somatic embryos after freezing in liquid nitrogen. *C. R. Acad. Sci. Paris*, **301**, Serie III, no. 3, 111–116.

Engelmann, F. and Duval, Y. (1986) Cryopreservation of oil palm somatic embryos (*Elaeis guineensis* jacq.): Results and application prospects. *Oleagineaux*, **41**(4), 169–174.

Engelmann, F., Duval, Y. and Pannetier, C. (1987) in *International Oil Palm/Palm Oil Conference: Progress and Prospects*, Kuala Lumpur, Malaysia. In press.

Evers, P. W., Donkers, J., Prat, A. and Vermeer, E. (1988). *Micropropagation of Forest Trees through Tissue Culture*, Pudoc, Wageningen.

FAO (1980) *1980 FAO Publication Yearbook*, FAO Statistics Series No. 34, Food and Agriculture Organisation of the United Nations, Rome.

FAO (1982) *1982 FAO Production Yearbook*, Food and Agricultural Organisation of the United Nations, Rome.

Favre, J. M. and Juncker, B. (1987) *In vitro* growth of buds taken from seedlings and adult plant material in *Quercus robur* L. *Plant Cell Tissue and Organ Culture*, **8**, 40–60.

Feucht, W. and Dausend, B. (1976) Root induction *in vitro* of easy-to-root Prunus pseudocerasus and difficult-to-root *P. avium*. *Sci. Hort.*, **4**, 49–54.

Finer, J. J., Kriebel, H. B. and Bocwar, M. R. (1989) Initiation of embryogenic callus and suspension cultures of eastern white pine (*Pinus strobus* L.). *Plant Cell Rpts*, **8**, 203–206.

Franclet, A. (1979) in *Micropropagation d'arbres forestiers*, AFOCEL – Etudes et Recherches no. 12, pp. 1–18.

Franclet, A. (1981) Rajeunissement et propagation vegetative des ligneux. *Ann. Rech. Silvicoles AFOCEL*, 12–40.

Fogle, H. W. and Winters, H. F. (1980) *North American and European Fruit and Tree Nut Germplasm Resources Inventory*, US Department of Agriculture Miscellaneous Publication No. 1406.

Frohlich, H. J. (1961) Untersuchungen über dasphysiologische und morphologische Verhalten von Vegetativvermehrungen verschiedener Laubund Nadelbaumarten. *Allg. Forst. u. J. Ztg.*, **132**, 39–58.

Galzy, R. (1969) Recherches sur la croissance de *Vitis rupestris* Scheele sain et court noice cultive *in vitro* a differentes temperatures. *Ann. Phytopathol.*, **1**, 149–166.

Ganeshan, S. (1985) Cryogenic preservation of grape (*Vitis unifera* L.) pollen. *Vitis*, **24**, 169–173.

Ganeshan, S. (1986) Cryogenic preservation of papaya pollen. *Sci. Hort.*, **28**, 65–70.

Ganeshan, S. and Sulladmath, V. V. (1983) Pollen storage studies on *Citrus limon* Burm. Varietal differences and influence of flower types. *Gartenbauwissenschaft*, **48**(2), 51–54.

Garner, R. J and Hatcher, E. S. (1962) Regeneration in relation to vegetative vigor and flowering. *Proc. XVI Int. Hort. Cong. Brussels*, **3**, 105–111.

Gautheret, R. J. (1934) Culture du tissue cambial. *C. R. Acad. Sci. (Paris)*, **198**, 2195–2196.

Gautheret, R. J. (1937) Nouvelles recherches sur la culture du tissue cambial. *C. R. Acad. Sci (Paris)*, **205**, 572–574.

Gautheret, R. J. (1940) Nouvelles recherches sur la bouregeonnement du tissu cambial d'*Ulmus campestris* cultive *in vitro*. *C. R. Acad. Sci. (Paris)*, **210**, 744–746.

Geissbühler, H. and Skoog, F. (1957) Comments on the application of plant tissue cultivation to propagation of forest trees. *Tappi*, **40**, 258–262.

Girouard, R. M. (1971) *Vegetative Propagation of Pines by Means of Needle Fascicles – A Literature Review*, Inf. Rep. Q-X-33, Canadian Forestry Service, Centre de Recherche Forestiere des Laurentides Quebec Region, Canada.

Girouard, R. M. (1974) Propagation of spruce by stem cuttings. *N. Z. J. For. Sci.*, **4**, 140–149.

Goyal, Y., Bingham, R. L. and Felker, P. (1985) Propagation of the tropical tree, *Leucaena leucocephala* K67, by *in vitro* bud culture. *Plant Cell Tissue and Organ Culture*, **4**, 3–10.

Grasser, J. W., Gmitter, F. G. Jr and Chandler, J. L. (1988) Intergeneric somatic hybrid plants of *Citrus sinensis* cf. Hamlin and *Poncirus trifoliata* cf. Flying Dragon. *Plant Cell Rpts*, **7**, 5–8.

Grout, B. W. W. (1986) in *Plant Tissue Culture and its Agricultural Applications* (ed. L. A. Withers and P. G. Alderson), Butterworths, London, pp. 303–310.

Grout, B. W. W., Shelton, K. and Pritchard, H. W. (1983) Orthodox behaviour of oil palm seed and cryopreservation of the excised embryo for genetic conservation. *Ann. Bot.*, **52**, 381–384.

Gulick, P. and van Sloten, D. H. (1984) *Directory of Germplasm Collections: Tropical and Subtropical Fruits and Tree Nuts*, IBPGR Secretariat, Rome.

Gupta, P. P. (1986) Eradication of mosaic disease and rapid clonal multiplication of both banana and plantains through meristem tip culture. *Plant Cell Tissue and Organ Culture*, **6**(1), 33–39.

Gupta, P. K., Nadgir, A. L., Mascarenhas, A. F. and Jogannathan, V. (1979), Tissue culture of forest trees: clonal multiplication of *Tectonia grandis* L. (teak) by tissue culture. *Plant Sci. Lett.*, **17**, 259–268.

Gupta, P. K. Mascarenhas, A. F. and Jagannathan, V. (1981) Tissue culture of forest trees – clonal propagation of mature trees of *Eucalyptus citriodora* Hook, by tissue culture. *Plant Sci. Lett.*, **20**, 195–201.

Haines, R. J. and De Fossard, R. A. (1977) Propagation of Hoop Pine (*Araucaria cuninghamii* Ait.) by organ culture. *Acta Hort.*, **78**, 297–301.

Haissig, B. E. (1965) Organ formation *in vitro* as applicable to forest tree propagation. *Bot. Rev.*, **31**, 607–626.

Hanover, J. W. and Keathley, D. E. (eds) (1988) *Genetic Manipulation of Woody Plants*, Plenum Press, New York.

Hanson, J. (1984) in *Crop Genetic Resources, Conservation and Evaluation* (ed. J. H. W. Holden and J. T. Williams), Allen and Unwin, London, pp. 53–62.

Hartney, V. J. (1980) Vegetative propagation of the Eucalyptus. *Aus. For. Res.*, **10**, 191–211.

Hasegawa, P. M. (1979) *In vitro* propagation of rose. *Hort. Sci.*, **14**, 610–612.

Hasegawa, P. M. (1980) Factors affecting shoot and root initiation from cultured rose shoot tips. *J. Am. Soc. Hort. Sci.*, **105**, 216–220.

Hatcher, E. S. J. and Garner, R. J. (1955) *The Production of Sphaeroblast Shoots of Apple for Cuttings*, Report of East Malling Research Station for 1954, Maidstone, Kent, pp. 73–75.

Hawkes, J. G. (1976) in *Tropical Trees, Variation, Breeding and Conservation* (ed. J. Burley and B. T. Styles), Academic Press, London.

Hawkes, J. G. (1982) in *Crop Genetic Resources – the Conservation of Difficult Material* (ed. L. A. Withers and J. T. Williams), IUBS Series, pp. 83–92.

Hidata, T. and Kajiura, I. (1988) Plantlet differentiation from callus protoplasts induced from *Citrus* embryo. *Sci Hort.*, **34**, 85–92.

Holden, J. H. W. (1984). in *Crop Genetic Resources; Conservation and Evaluation* (ed. J. H. W. Holden and J. T. Williams), George Allen and Unwin, London, pp. 117–185.

Howard, H. B. (1978) *Propagation and Nursery Production*, Report of East Malling Research Station for 1977, Maidstone, Kent, pp. 67–71.

Huang, S. C. and Millikan, D. F. (1980) *In vitro* micrografting of apple shoot tips. *Hort. Sci.*, **15**(6), 741–743.

Hudson, J. P. (1982) New perspectives in vegetative propagation. *Outlook on Agriculture*, **11**(2), 55–61.

Hutchinson, J. F. (1982) in *Tissue Culture of Economically Important Plants* (ed. A. N. Rao), COSTED, Singapore, pp. 113–120.

Hwang, S. C., Chen, C. L., Lin, J. C. and Lin, H. L. (1984) Cultivation of banana using plantlets from meristem culture. *Hort. Sci.*, **119**, 231–233.

IBPGR (1983) *Annual Report for 1982*, IBPGR Executive Secretariat, Rome.

IBPGR (1985) *Long-Term Seed Storage of Major Temperate Fruits*, IBPGR, Rome.

IBPGR (1986a) *Genetic Resources of Tropical and Sub-tropical Fruits and Nuts (Excluding Musa)*, IBPGR Secretariat, Rome.

IBPGR (1986b) *IBPGR Advisory Committee on* In Vitro *Storage: Design, Planning and Operation of* In Vitro *Genebanks. Report of a Subcommittee Meeting*, IBPGR, Rome.

Ishihara, A., Obonai, Y., Ito, Y. and Kobayashi, S. (1987) Experiments for cryopreservation of apple shoot tips isolated from shoots culltured *in vitro. Jpn. Soc. Hort. Sci. Abstracts*, 248–249.

IUCN (1978) *The IUCN Plant Red Data Book* (ed. G. Lucas and H. Synge), IUCN, Switzerland.

Iyer, R. D. (1982) in *Tissue Culture of Economically Important Plants* (ed. A. N. Rao), COSTED, Singapore, pp. 219–230.

Jacquiot, C. (1949) Observations sur la neoformation de bourgeons chez le tissu cambial d'*Ulmus campestris* cultive *in vitro. C. R. Acad. Sci. (Paris)*, **229**, 529–530.

Jacquiot, C. (1951) Action du mesoinositol et de l'adenine sur la formation de bourgeons par le tissue cambial d'*Ulmus campestris* cultive *in vitro. C. R. Acad. Sci. (Paris)*, **233**, 815–817.

Jacquiot, C. (1955a) Formation d'organes par le tissue cambial d'*Ulmus campestris* L. et de *Betula verrucosa* Gaertn. cultives *in vitro. C. R. Acad. Sci. (Paris)*, **240**, 557–558.

Jacquiot, C. (1955b) Sur le role des correlations d'inhibition dan les phenomenes d'organogenese observes chez le tissue cambial, cultive *in vitro*, de certains arbes. Incidences sur les problems du bouturage. *C. R. Acad. Sci. (Paris)*, **240**, 1064–1066.

James, D. J. and Thurbon, I. J. (1979) Rapid *in vitro* rooting of the apple rootstock M9. *J. Hort. Sci.*, **54**(4), 309–311.

James, D. J., Plessey, A. J., Barbara, D. J. and Bevan, M. (1989) Genetic transformation of apple (*Malus pumila* Mill.) using a disarmed Ti-binary vector. *Plant Cell Rpts*, **7**, 658–661.

Janick, J. (1979) Horticulture's ancient roots. *Hort. Sci.*, **14**, 299–313.

Jarret, R. L. (1986) in *IBPGR Advisory Committee on* In Vitro *Storage: Report of the Third Meeting*, IBPGR, Rome, pp. 15–30.

References

Jarret, R. L. and Litz, R. E. (1986) Isozymes as genetic markers in bananas and plantains. *Euphytica*, **35**, 539–549.

John, A. (1983) in *Tissue Culture of Trees* (ed. J. H. Dodds), Croom-Helm, London.

Jones, L. H. (1974) Propagation of clonal oil palms by tissue culture. *Oil Palm News*, **17**, 1–8.

Jones, O. P. (1967) Effect of benzyladenine on isolated apple shoots. *Nature*, **215**, 1514–1515.

Jones, O. P. (1978) in *Round Table Conference*, CRA Gembloux, Belgium, p. 22.

Jones, O. P. (1979) Propagation *in vitro* of apple trees and other woody fruit plants. *Sci. Hort.*, **30**, 44–48.

Jones, O. P. and Hopgood, M. E. (1979) The successful propagation *in vitro* of two rootstocks of *Prunus*: the plum rootstock Pixy (*P. insititia*) and the cherry rootstock F12/1 (*P. avium*). *J. Hort. Sci.*, **54**, 63–66.

Jones, O. P. and Vine, S. J. (1968) The culture of gooseberry shoot tips for eliminating virus. *J. Hort. Sci.*, **43**, 289–292.

Jones, O. P., Hopgood, M. E. and O'Farrell, D. (1977) Propagation *in vitro* of M26 apple rootstocks. *J. Hort. Sci.*, **52**, 235–238.

Jones, O. P., Pontikis, C. A. and Hopgood, M. E. (1979) Propagation *in vitro* of five apple scion cultivars. *J. Hort. Sci.*, **54**, 155–158.

Kajiura, I. (1980) Kaki culture in Japan (the persimmon). *Orchardist N. Zealand*, **53**, 98–100.

Karnosky, D. F. (1981) Potential for forest tree improvement via tissue culture. *BioScience*, **31**, 144–120.

Kartha, K. K. (1985) *Cryopreservation of Plant Cells and Organs*, CRC Press, Boca Raton, Florida.

Kartha, K. K., Mroginski, L. A., Pahl, K. and Leung, N. L. (1981) Germplasm preservation of coffee (*Coffea arabica* L.) by *in vitro* culture of shoot apical meristems. *Plant Sci. Lett.*, **22**, 301–307.

Kartha, K. K., Fowke, L. C., Leung, N. L., Caswell, K. L. and Haman, I. (1989) Induction of somatic embryos and plantlets from cryopreserved cell cultures of White Spruce (*Picea glauca*). *J. Plant Physiol.*, **132**, 529–539.

Katano, M., Ishihara, A. and Sakai, A. (1983) Survival of dormant apple shoot tips after immersion in liquid nitrogen. *Hort. Sci.*, **18**(5), 707–708.

Katano, M., Ishihara, A. and Sakai, A. (1984) Survival of apple shoot tips isolated from shoots in culture after immersion in liquid nitrogen. *Jpn. J. Breeding*, **34** (Suppl. 1), 212–213.

Keays, J. L. (1974) Full-tree and complete-tree utilization for pulp and paper. *For. Prod. J.*, **24**, 13–16.

Kevers, C. and Gasper, Th. (1985) Soluble membrane and wall peroxidases, phenylalanine ammonia-lyase and lignin changes in relation to nitrification of carnation tissues cultured *in vitro*. *J. Plant. Physiol.*, **118**, 41–48.

King, M. W. and Roberts, E. H. (1979a) The desiccation response of seeds of *Citrus limon* L. *Ann. Bot.*, **45**, 489–492.

King, M. W. and Roberts, E. H. (1979b) *The Storage of Recalcitrant Seeds – Achievements and Possible Approaches*, International Board for Plant Genetic Resources, Rome.

King, M. W. and Roberts, E. H. (1980) in *Recalcitrant Crop Seeds* (ed. H. F. Chin and E. H. Roberts), Tropical Press, Kuala Lumpur, Malaysia, pp. 90–110.

Kobayashi, S., Ikeda, I. and Nakatani, M. (1978) in *Long Term Preservation of Favourable Germ Plasm in Arboreal Crops* (ed. T. Akihama and K. Nakajima), Fruit Tree Research Station, M.A.F. Japan, pp. 8–12.

Kobayashi, S., Ishihara, A. and Sakai, A. (1987) in *Abstracts of 10th Symposium for Plant Tissue Culture*, Sendai, Japan. In press.

Kong, L. S. and Rao, A. N. (1982) in *Tissue Culture of Economically Important Plants* (ed. A. N. Rao), COSTED, Singapore, pp. 185–190.

Kormanik, P. P. and Brown, C. L. (1974) Vegetative propagation of some selected hardwood forest species in the south-eastern United States. *N. Z. J. For. Sci.*, **4**, 228–234.

Kozlowski, T. T. (1971) *Growth and Development of Trees*, Vol. 1, Academic Press, New York, pp. 94 and 117.

Krul, W. R. and Mowbury, G. H. (1984) in *Handbook of Plant Cell Culture*, Vol. 2, *Crop Species* (ed. W. A. Sharp, D. A. Evans, P. V. Ammirato and Y. Yamada), MacMillan, London, pp. 396–434.

Kyte, L. and Briggs, B. (1979) A simplified entry into tissue culture propagation of rhododendrons. *Comb. Proc. Int. Plant Prop. Soc.*, **29**, 90–95.

Lakshmi-Sita, G., Chattopadhyay, S. and Tejavathi, D. H. (1986) Plant regeneration from shoot callus of rosewood (*Dalbergia latifolia* Roxb.). *Plant Cell Rpts*, **5**, 266–268.

Lamb, R. C. (1974) Future germplasm reserves of pome fruits. *Fruit Var. J.*, **28**, 75–79.

Lane, W. D. (1979) Regeneration of pear plants from shoot meristem tips. *Plant Sci. Lett.*, **16**, 337–342.

Lane, W. D. (1982) in *Application of Plant Cell and Tissue Culture to Agriculture and Industry* (ed. D. T. Tomes, B. E. Ellis, P. M. Harney, K. J. Kasha and R. L. Peterson), University of Guelph, Ontario, pp. 163–186.

Lang, H. and Kohlenbach, H. W. (1988) Callus formation from mesophyll protoplasts of *Fagus sylvatica* L. *Plant Cell Rpts*, **7**, 485–488.

Lanly, J. P. (1982) *Tropical Forest Resources*, FAO Forestry Paper No. 30, FAO, Rome, Italy.

Lee, S. K. and Rao, A. N. (1986) *In vitro* regeneration of plantlets in *Fagraea fragrans* Roxb. – a tropical tree. *Plant Cell Tissue and Organ Culture*, **7**, 43–51.

Libby, W. J. (1974) The use of vegetative propagules in forest genetics and tree improvement. *N. Z. J. For. Sci.*, **4**, 440–447.

Libby, W. J., Strittler, R. F. and Seitz, F. W. (1969) Forest genetics and forest-tree breeding. *Annu. Rev. Genet.*, **3**, 469.

Litz, R. E. (1985) in *Tissue Culture in Forestry and Agriculture* (ed. R. R. Henke, K. W. Hughes, M. J. Constantin and A. Hollaender), Plenum Press, New York, pp. 179–193.

Litz, R. E., Knight, R. J. and Gazit, S. (1982) Somatic embryos from cultured ovules of polyembryonic *Mangifera indica* L. *Plant Cell Rpts,* **1**, 264–266.

Litz, R. E., Knight, R. J. and Gazit, S. (1984) *In vitro* somatic embryogenesis from *Mangiera indica* L. callus. *Sci. Hort.*, **22**, 233–240.

Longman, K. A. (1976) in *Tropical Trees, Variation, Breeding and Conservation* (ed. J. Burley and B. T. Styles), Academic Press, London, pp. 19–24.

Lopez, M. M. and Navarro, L. (1981) in *Proceedings of the International Society of Citriculture, 1981*, Vol. 1, pp. 399–401.

Loreti, F. and Morini, S. (1982) Mass propagation of fruit trees in Italy by tissue culture: Present status and perspectives. *Comb. Proc. Int. Plant Prop. Soc.*, **32**, 283–291.

Lugo, A. E. (1985) in *Tissue Culture in Forestry and Agriculture* (ed. R. R. Henke, K. W. Hughes, M. J. Constantin and A. Hollaender), Plenum Press, New York, pp. 289–295.

Lundergan, C. and Janick, J. (1979) Low temperature storage of *in vitro* apple shoots. *Hort. Sci.*, **14**, 514.

Lyr, H., Polster, H. and Fiedler, H. J. (1967) *Geholzphysiologie*, Gustav Fisher Verlag, Jena.

Lyrene, P. M. (1981) Juvenility and production of fast-rooting cuttings from blueberry shoot cultures. *J. Am. Soc. Hort. Sci.*, **106**, 396–398.

Ma, S. and Shii, C. (1972) *In vitro* formation of adventitious buds in banana shoot apex following decapitation. *J. Hort. Sci. China*, **18**, 135–142.

Ma, S. and Shii, C. (1974) Growing bananas from adventitious buds. *J. Hort. Sci. China*, **20**, 6–12.

Mackay, J., Seguin, A. and Lalonde, M. (1988) Genetic transformation of 9 *in vitro* clones of *Alnus* and *Betula* by *Agrobacterium tumefaciens*. *Plant Cell Rpts*, **7**, 229–232.

Maggs, D. H. (1973) Genetic resources in pistachio. *Plant Genet. Res. Newsletter*, **29**, 7–15.

Maggs, D. H. and Alexander, D. McE (1967) A topophysic relation between regrowth and pruning in *Eucalyptus cladocalyx* F. Muell. *Aust. J. Bot.*, **15**, 1–9.

Marino, G., Rosati, P. and Sagrati, F. (1985) Storage of *in vitro* cultures of *Prunus* rootstocks. *Plant Cell Tissue and Organ Culture*, **5**(1), 73–78.

Marshall, D. R. (1989) in *The Use of Plant Genetic Resources* (ed. A. H. D. Brown, O. H. Frankel, D. R. Marshall and J. T. Williams), Cambridge University Press, Cambridge, pp. 105–120.

Martin, C., Carre, M. and Vernoy, R. (1981) La multiplication vegetative *in vitro* des vegetaux ligneux cultives: Cas de rosiers. *C. R. Acad. Sci. (Paris)*, Ser. III, **293**, 175–177.

Mascarenhas, A. F., Gupta, P. K., Kulkarni, V. M., Mehta, U., Iyer, R. S., Khuspe, S. S. and Jagannathan, V. (1982) in *Tissue Culture of Economically Important Plants* (ed. A. N. Rao), COSTED, Singapore, pp. 175–179.

Mathes, M. C. (1964) The *in vitro* formation of plantlets from aspen tissue. *Phyton*, **21**, 137–141.

McAlpine, R. G. (1964) A method of producing clones of yellow poplar. *J. For.*, **62**, 115–116.

Meins, J. F. and Binns, A. N. (1979) Cell determination in plant development. *BioScience*, **29**, 221–225.

Meryman, H. T. and Williams, R. J. (1985) in *Cryopreservation of Plant Cells and Organs* (ed. K. Kartha), CRC Press, Boca Raton, Florida.

Miller, G. A., Coston, D. C., Denny, E. G. and Romero, M. E. (1982) *In vitro* propagation of 'Nemaguard' peach rootstock. *Hort. Sci.*, **17**(2), 194.

Monette, P. L. (1986) Cold storage of Kiwifruit shoot tips *in vitro*. *Hort. Sci.*, **21**(5), 1203–1205.

Monteuwis, O. (1986) *In vitro* micrografting of *Sequoiadendron giganteum* meristems. *C. R. Acad. Sci. (Paris)*, **302**, Serie III, no. 6.

Moriguchi, T., Akihama, T. and Kozaki, I. (1985) Freeze-preservation of dormant pear shoot apices. *Jpn. J. Breeding*, **35**, 196–199.

Mott, R. L. (1981) in *Cloning Agricultural Plants via in vitro Techniques* (ed. B. V. Conger), CRC Press, Boca Raton, Florida, pp. 217–256.

Mukherjee, S. K. (1985) *Systemic and Ecogeographic Studies in Crop Genepools: I. Mangifera L.*, International Board for Plant Genetic Resources, Rome.

Mullins, M. G. and Srinivasan, C. (1976) Somatic embryos and plantlets from an ancient clone of grapevine (cv. 'Cabernet-Sauvignon') by apomixis *in vitro*. *J. Exp. Bot.*, **27**, 1022–1030.

Mullins, M. G., Nair, Y. and Sampet, P. (1979) Rejuvenation *in vitro*: introduction of juvenile characters in an adult clone of *Vitis vinifera* L. *Ann. Bot.*, **44**, 623–627.

National Academy of Sciences (1975) *Underexploited Tropical Plants with Promising Economic Value*, Report of an Advisory Committee on Technical Innovation, National Academy of Sciences, Washington DC.

National Academy of Sciences (1979) *Tropical Legumes: Resources for the Future*, National Academy of Sciences, Washington DC.

National Academy of Sciences (1984) *Casuarina: Nitrogen Fixing Trees for Adverse Sites*, American National Academy of Sciences, Washington DC.

Navarro, L., Roistacher, C. N. and Murashige, T. (1975) Improvement of shoot tip grafting *in vitro* for virus free Citrus. *J. Am. Soc. Hort. Sci.*, **100**, 471–479.

Nepveu, G. (1982) Variabilite clonal de l'infadensite chez *Quercus petraea*. Premiers resultats obtenus sur boutures d'un an. *Ann. Sci. Forest.*, **32**(2), 151–164.

Nepveu, G. (1984) Determinisme genotypique de la structure anatomique du bois chez *Quercus robur*. *Silvae Genetica*, **33**(2–3), 91–94.

Noiret, J. M., Gascon, J. P. and Pannetier, C. (1985) Oil-palm production through *in vitro* culture. *Oleagineux*, **40**, 365–372.

Nwankwo, B. A. and Krikorian, A. D. (1986) Morphogenetic potential of embryo- and seedling-derived callus of *Elaeis guineensis* Jacq. var. pissifera Becc. *Ann. Bot.*, **51**, 65–76.

Ochatt, S. J., Cocking, E. C. and Power, J. B. (1987) Isolation, culture and plant regeneration of colt cherry (*Prunus avium* × *pseudocerasus*) protoplasts. *Plant Sci.*, **50**, 139–143.

Ochatt, S. J., Rech., Davey, M. R. and Power, J. B. (1988) Long term effect of electroporation on enhancement of growth and plant regeneration of colt cherry (*Prunus avium* × *pseudocerasus*) protoplasts. *Plant Cell Rpts*, **7**, 393–395.

Ohgawara, T., Kobayashi, S., Ohgawara, E., Uchimiya, H. and Ishii, S. (1985) Somatic hybrid plants obtained by protoplast fusion between *Citrus sinensis* and *Poncirus trifoliata*. *Theor. Appl. Genet.*, **71**, 1–4.

Paily, J. and D'Souza, L. (1986) *In vitro* clonal propagation of *Lagerstroemia flos-reginae* Retz. *Plant Cell Tissue and Organ Culture*, **6**(1), 41–45.

Palmberg, C. (1984) in *Crop Genetic Resources: Conservation and Evaluation* ed. J. H. W. Holden and J. T. Williams), George Allen and Unwin, London, pp. 223–237.

Pannetier, C., Arthuis, P. and Lievoux, D. (1981) Neo-formation of young *E. guineensis* plantlets from primary calluses obtained on leaf fragments cultured *in vitro*. *Oleagineaux*, **36**(3), 119–122.

Parfitt, D. E., Aruselkar, S. and Ramming, D. W. (1985) Identification of plum × peach hybrids by isozyme analysis. *Hort. Sci.*, **20**, 246–248.

Passecker, F. (1964) Lassen sich gealterte Obstsorten verjungen? *Mitteilungen serie b Obst. und Garten*, **14**, 196–200.

Peacock, W. J. (1989) in *The Use of Plant Genetic Resources* (ed. A. H. D. Brown, O. H. Frankel, D. R. Marshall and J. T. Williams), Cambridge University Press, Cambridge, pp. 363–376.

Pearce, F. (1989) Kill or cure? Remedies for the rainforest. *New Scientist*, **123**, 40–43.

Perron, G. (1981) L'Eucalyptus a la conquete de l'Europe. *PHM. Rev. Hort.*, **214**, 13–16.

Peters, J. P. and Williams, J. T. (1984) Towards better use of genebanks with special reference to information. *Plant Genet. Res. Newsletter*, **60**, 22–31.

Pieriazek, J. (1968) The growth *in vitro* of isolated apple-shoots from young seedlings on media containing growth regulators. *Bull. Acad. Pol. Sci. Ser. Sci. Biol.*, **16**, 179–183.

Poissonnier, M., Franclet, A., Dumant, M. J. and Gautry, J. Y. (1980) *Ann. AFOCEL*, 232–253.

Pritchard, H. W. and Prendergast, F. G. (1986) Effects of desiccation and cryopreservation on the *in vitro* viability of embryos of the recalcitrant seed species *Araucaria husteinii* K. Schum. *J. Exp. Bot.*, **37**(182), 1388–1397.

Pritchard, H. W., Grout, B. W. W., Reid, D. S. and Short, K. C. (1982) in *Biophysics of water* (ed. F. Franks and S. F. Mathias), John Wiley, Chichester, pp. 315–318.

Quoirin, M. (1974) Premiers resultats obtenus dans la culture *in vitro* du meristeme apical du sujets porte-greffe due pommier. *Bull. Rech. Agronom. Gembloux*, **9**, 189–192.

Rabechault, H. and Martin, J.-P. (1976) Multiplication vegetative du Palmier a huile (*Elaeis guineensis* Jacq.) a l'aide de cultures de de tissus foliares. *C. R. Acad. Sci. Ser. III Vie*, **283**, 1735–1737.

Rao, A. N. and Lee, S. K. (1986) in *Plant Tissue Culture and its Agricultural Applications* (ed. L. A. Withers and P. G. Alderson), Butterworths, London, pp. 123–138.

Raven, P. H. (1980) *Conversion of Tropical Moist Forests*, NRC Committee Report on Tropical Biology, National Academy of Sciences, Washington DC.

Redenbaugh, K., Paasch, B. D., Nichol, J. W., Kossler, M. E., Viss, P. R. and Walker, K. A. (1986) Somatic seeds: encapsulation of asexual plant embryos. *Biotechnology*, **4**, 797–801.

Reed, B. M. and Lagerstedt, H. B. (1987) Freeze preservation of apical meristems of *Rubus* in liquid nitrogen. *Hort. Sci.*, **22**(2), 302–303.

Reisch, B. (1984) in *Handbook of Plant Cell Culture, Techniques for Propagation*, Vol. 1 (ed. D. A. Evans, W. R. Sharp, P. V. Ammirato and Y. Yamada), MacMillan, New York, pp. 748–769.

Reuveni, O., Israeli, Y., Degani, H. and Eshdat, Y. (1985) *Genetic Variability in Banana Plants Multiplied via in vitro Techniques*, IBPGR Internal Report/ 85/9, IBPGR.

Reynolds, J. F. (1982) in *Tissue Culture in Forestry* (ed. J. M. Bonga and D. J. Durzan), Martinus Nijhoff, London, pp. 182–207.

Reynolds, J. F. and Murashige, T. (1979) Asexual embryogenesis in callus cultures of palms. *In Vitro*, **15**, 383–387.

Robbins, W. J. (1964) Topophysics, a problem in somatic inheritance. *Proc. Am. Philos. Soc.*, **108**, 395.

References

Roberts, E. H. and King, M. W. (1981) in *Crop Genetic Resources – the Conservation of Difficult Material* (ed. L. A. Withers and J. T. Williams), IUBS Series B42, pp. 39–48.

Rogers, W. S. and Beakbane, A. B. (1957) Stock and scion relations. *Annu. Rev. Plant Physiol.*, **8**, 217–236.

Rosati, P., Marino, G. and Swierczewski, C. (1980) *In vitro* propagation of Japanese plum (*Prunus salicina* Lincl. cv. Calita). *J. Am. Soc. Hort. Sci.*, **105**, 126–129.

Ross, S. D. (1975) Production, propagation and shoot elongation of cuttings from sheared 1-year old Douglas-fir seedlings. *For. Sci.*, **21**, 298–300.

Rowe, P. (1984) Breeding bananas and plantains. *Plant Breed. Rev.*, **2**, 135–155.

Rubeli, K. (1989) Pride and protest in Malaysia. *New Scientist*, **124**, 49–52.

Rugini, E. and Verma, D. C. (1983) Micropropagation of difficult to propagate almond (*Prunus amygdalus*) cultivar. *Plant Sci. Lett.*, **28**, 273–281.

Russell, J. A. and McCown, B. H. (1988) Recovery of plants from leaf protoplasts of hybrid poplar and aspen clones. *Plant Cell Rpts*, **7**, 67–69.

Sakai, A. (1985) in *Cryopreservation of Plant Cells and Organs* (ed. K. Kartha), CRC Press, Boca Raton, Florida.

Sakai, A. and Nishiyama, Y. (1978) Cryopreservation of winter vegetative buds of hardy fruit trees in liquid nitrogen. *Hort. Sci.*, **13**, 225–227.

Sakai, A. and Sugawara, Y. (1973) Survival of poplar callus at super-low temperatures after cold acclimation. *Plant Cell Physiol.*, **14**, 1201–1204.

Sastrapradja, S. (1975) in *East Asian Plant Genetic Resources* (ed. J. T. Williams, C. H. Lamoureux and N. Wulijarni-Soetjipto), IBPGR/SEAMEO/BIOTROP/LIPI, Bogor, pp. 33–46.

Sattaur, O. (1983) High-speed tree saves Timor's soil. *New Scientist*, **99**, 104.

Schopke, C., Miller, L. E. and Kohlenbach, H-W. (1987) Somatic embryogenesis and regeneration of plantlets in protoplast culturs from somatic embryos of coffee (*Coffea canephora* P. ex Fr.). *Plant Cell Tissue and Organ Culture*, **8**, 243–248.

Scowcroft, W. R. (1984) *Genetic Variability in Tissue Culture: Impact on Germplasm Conservation and Utilisation*, IBPGR. Rome.

Seguin, A. and Lalonde, M. (1988) Gene transfer by electroporation in betulaceae protoplasts: *Alnus incana*. *Plant Cell Rpts*, **7**, 367–370.

Sekowski, B. (1956) *Pomologia*, Poznan.

Shields, W. J. Jr and Bockheim, J. G. (1981) Deterioration of trembling aspen clones in the Great Lakes region. *Can. J. For. Res.*, **1**, 530–537.

Simmonds, N. W. (1982) in *Crop Genetic Resources – the Conservation of Difficult Material* (ed. L. A. Withers and J. T. Williams), IUBS Series, pp. 1–3.

Simpson, M. J. A. and Withers, L. A. (1986) *Characterisation Using Isozyme Electrophoresis: A Guide to the Literature*, Rome, IBPGR.

Singha, S. (1980) in *Proc. Conf. Nursery Production of Fruit Plants through Tissue Culture – Applications and Feasibility*, US Dept. Agr., SEA. ARR-NE-11, pp. 59–63.

Singha, S. (1982) *In vitro* propagation of crabapple cultivars. *Hort. Sci.*, **17**(2), 191–192.

Skirvin, R. M. (1981) *Cloning Agricultural Plants via In vitro Techniques* (ed. B. V. Conger), CRC Press, Boca Raton, Florida.

Skirvin, R. M. and Chu, M. C. (1979) *In Vitro* propagation of 'Forever Yours' rose. *Hort. Sci.*, **14**, 608–610.

Smith, D. R., Horgan, K. J. and Aitken-Christie, J. (1982) Micropropagation of *Pinus radiata* for afforestation, in *Proc. 5th Int. Cong. Plant Tissue and Cell Culture*, (ed. A. Fujiwara) Japanese Assoc. for Plant Tissue Culture, Tokyo, pp. 723–724.

Snir, I. (1982) *In vitro* propagation of sweet cherry cultivars. *Hort. Sci.*, **17**(2), 192–193.

Sommer, H. E., Brown, C. L. and Kormanik, P. P. (1975) Differentiation of plantlets in longleaf pine (*Pinus pallustris* Mill.) tissue cultured *in vitro*. *Bot. Gaz.*, **136**, 196–200.

Sondahl, M. R., Nakamura, T. and Sharp, W. R. (1985) in *Tissue Culture in Forestry and Agriculture* (ed. R. R. Henke, K. W. Hughes, M. J. Constantin and A. Hollaender), Plenum Press, New York, pp. 215–232.

Sossou, J., Karaunaratne, S. and Kouoor, A. (1987) Collecting palm: *in vitro* explanting in the field. *FAO/IBPGR Plant Genet. Resources Newsletter*, **69**, 7–18.

Sriskandarajah, S., Mullins, M. G. and Nair, Y. (1982) Induction of adventitious rooting *in vitro* in difficult to propagate cultivars of apple. *Plant Sci. Lett.*, **18**, 1–9.

Stushnoff, C. (1987) Cryopreservation of apple genetic resources. *Can. J. Plant Sci.* (in press).

Stushnoff, C. and Fear, C. (1985) *The Potential Use of in vitro Storage for Temperate Fruit Germplasm: A Status Report*, IBPGR, Rome.

Sugawara, Y. and Sakai, A. (1974) Survival of suspension cultured sycamore cells cooled to the temperature of liquid nitrogen. *Plant Physiol.*, **54**, 722–724.

Sutter, E. G. and Barker, P. B. (1985). *In vitro* propagation of mature *Liquidambar styraciflua*. *Plant Cell Tissue and Organ Culture*, **5**(1), 13–21.

Suttle, G. R. L. (1983) Micropropagation of deciduous trees. *Int. Plant Prop. Soc. Comb Proc.*, **33**, 46–49.

Sykes, J. T. (1975) in *Crop Genetic Resources for Today and Tomorrow* (ed. O. H. Frankel and J. G. Hawkes), Cambridge University Press, Cambridge, pp. 123–127.

References

Tanksley, S. D. and Orton, T. J. (1983) *Isozymes in Plant Genetics and Breeding*, Part A and B, Elsevier, Amsterdam.

Thompson, M. M. (1981) Utilisation of fruit and nut germplasm. *Hort. Sci.*, **16**(2), 6–9.

Thorpe, T. A. and Biondi, S. (1984) in *Handbook of Plant Cell Culture*, Vol. 2, *Crop Species* (ed. W. R. Sharp, D. A. Evans, P. V. Ammirato and Y. Yamada), MacMillan, New York, pp. 435–470.

Thulin, I. J. and Faulds, T. (1968) The use of cuttings in the breeding and afforestation of *Pinus radiata*. *N. Z. J. For.*, **13**, 66–77.

Tisserat, B. (1979) Propagation of date palm (*Phoenix dactylifera* L.) *in vitro*. *J. Exp. Bot.*, **30**, 1275–1283.

Tisserat, B. (1982) Factors involved in the production of plantlets from date palm callus cultures. *Euphytica*, **31**, 201–214.

Tisserat, B. (1984) in *Handbook of Plant Cell Culture*, Vol. 2, *Crop Species* (ed. W. R. Sharp, D. A. Evans, P. V. Ammirato and Y. Yamada), MacMillan, New York, pp. 505–545.

Tisserat, B., Ulrich, J. M. and Finkle, B. J. (1981) Cryogenic preservation and regeneration of date palm tissue. *Hort. Sci.*, **16**, 47–48.

Tubbs, F. R. (1967) Tree size control through dwarfing rootstocks. *Proc. XVII Int. Hort. Cong.*, **3**, 43–56.

Ulrich, J. M., Finkle, B. J. and Tisserat, B. H. (1982) Effects of cryogenic treatment in plantlet production from frozen and unfrozen date palm callus. *Plant Physiol.*, **69**, 624–627.

Ulrich, J. M., Mickler, R. A., Finkle, B. J. and Karnosky, D. F. (1984) Survival and regeneration of American elm callus cultures after being frozen in liquid nitrogen. *Can. J. For. Res.*, **14**, 750–753.

Vardi, A., Spiegel-Roy, P. and Galun, E. (1975) *Citrus* cell culture: isolation of protoplasts, plating densities, effect of mutagens and regeneration of embryoids. *Plant Sci. Lett.*, **4**, 231–236.

Vardi, A., Spiegel-Roy, P. and Galun, E. (1982) Plant regeneration from *Citrus* protoplasts: Variation in methodological requirements for various species. *Theor. Appl. Genet.*, **62**, 171–176.

Vavilov, N. I. (1926) *Studies on the Origin of Cultivated Plants*, Leningrad.

Vavilov, N. W. (1930) Wild progenitors of the fruit trees of Turkistan and the Caucasus and the problem of origin of fruit trees. *Proc. 9th Int. Hort. Cong.*, 271–278.

Vertesy, J., Jones, O. P. and Hopgood, M. E. (1980) *Report of the East Malling Research Station for 1979*, Maidstone, Kent, p. 187.

Vivekanandan, K. (1978) Retention of viability of mahogany seed through cold storage. *Sri Lanka Forester*, **13**, 67–68.

Vuylsteke, D. and de Langhe, E. (1985) Feasibility of *in vitro* propagation of bananas and plantains. *Trop. Agri. (Trin.)*, **62**, 323–328.

Wakisaka, I. (1964) Ultra low temperature storage of pollens of Japanese persimmons. *J. Jpn. Soc. Hort. Sci.*, **33**, 291–294.

Walkey, D. G. (1972) Production of apple plantlets from axillary bud meristems. *Can. J. Plant Sci.*, **52**, 1085–1087.

Wan Abdul Rahaman, W. Y., Ghandimathi, H., Rohani, O. and Paranjothy, K. (1982) in *Tissue Culture of Economically Important Plants* (ed. A. M. Rao), COSTED, Singapore, pp. 151–158.

Wanas, W. H., Callow, J. C. and Withers, L. A. (1986) in *Plant Tissue Culture and its Agricultural Applications* (ed. L. A. Withers and P. G. Alderson), Butterworths, London, pp. 285–290.

Wareing, P. F. (1971) Some aspects of differentiation in plants. *Symp. Soc. Exp. Biol.*, **24**, 323–344.

Welander, M. (1985) *In vitro* shoot and root formation in the apple cultivar Akero. *Ann Bot.*, **55**, 249–261.

Wellensiek, S. J. (1952) Rejuvenation of woody plants by formation of spaeroblasts. *Proc. Koninklijke Nederlandse Akademie van Wetenschappen. Series C. Biological and Medical Sciences*, **55**, 567–573.

Westwood, M. N. (1978) *Temperate Zone Pomology*, W. H. Freeman and Company, San Francisco.

Wilkins, C. P. and Dodds, J. H. (1983a) in *Tissue Culture of Trees* (ed. J. H. Dodds), London, Croom Helm, pp. 113–136.

Wilkins, C. P. and Dodds, J. H. (1983b) in *Tissue Culture of Trees* (ed. J. H. Dodds), Croom Helm, London, pp. 56–79.

Wilkins, C. P. and Dodds, J. H. (1983c) The application of tissue culture techniques to plant genetic conservation. *Sci. Prog. (Oxf.)*, **63**, 259–284.

Wilkins, C. P., Cabrera, J. L. and Dodds, J. H. (1985) Tissue culture propagation of trees. *Outlook Agric.*, **14**, 2–13.

Wilkins, C. P., Newbury, H. J. and Dodds, J. H. (1988) Tissue culture conservation of fruit trees. *FAO/IBPGR Plant Genet. Resources Newsletter*, **73/74**.

Williams, J. T. (1982) in *Crop Genetic Resources – the Conservation of Difficult Material* (ed. L. A. Withers and J. T. Williams), IUBS Series, pp. 115–118.

Winton, L. (1970) Shoot and tree production from aspen tissue culture. *Am. J. Bot.*, **57**, 904–909.

Winton, L. (1971) Tissue culture propagation of European aspen. *For. Sci.*, **17**, 348–350.

Withers, L. A. (1978) The freeze-preservation of synchronously dividing cultured cells of *Acer pseudoplatanus* L. *Cryobiology*, **15**, 87–92.

Withers, L. A. (1980) *Tissue Culture Storage for Genetic Conservation*, IBPGR Technical Report, IBPGR Secretariat, Rome.

Withers, L. A. (1984) in *Crop Genetic Resources: Conservation and Evaluation* (ed. J. H. W. Holden and J. T. Williams), George Allen and Unwin, London, pp. 138–153.

Withers, L. A. (1987) *In vitro* methods for collecting germplasm in the field. *FAO/IBPGR Plant Genet. Resources Newsletter*, **69**, 2–6.

Withers, L. A. (1989) in *Use of Plant Genetic Resources* (ed. A. H. D. Brown, O. H. Frankel, D. R. Marshall and J. T. Williams), Cambridge University Press, Cambridge, pp. 309–334.

Withers, L. A. and Wheelans, S. K. (1986) in *IBPGR Advisory Committee on In Vitro Storage*, Report of the Third Meeting, IBPGR, Rome, pp. 13–14.

Wolter, K. E. (1968) Root and shoot initation in aspen callus cultures. *Nature*, **219**, 509–510.

Yidana, J. A., Withers, L. A. and Ivins, J. D. (1987) Development of a simple method for collecting and propagating cocoa germplasm *in vitro*. *Acta Hort.* (in press).

Yie, S. T. and Liaw, S. I. (1977) Plant regeneration from shoot tips and callus of papaya. *In vitro*, **13**, 564–568.

Zagaja, S. W. (1970) in *Genetic Resources in Plants – their Exploration and Conservation* (ed. O. H. Frankel and E. Bennet), IBP Handbook No. 11, Blackwell Scientific Publications, Oxford, pp. 327–333.

Zagaja, S. W. (1983) in *Methods in Fruit Breeding* (ed. J. N. Moore and J. Janick), Purdue University Press, West Lafayette, Indiana, pp. 3–10.

Zielinski, Q. B. (1955) *Modern Systematic Pomology*, Iowa State University Press, Ames, Iowa.

Zimmerman, R. H. (1978) Tissue culture of fruit trees and other fruit plants. *Proc. Int. Plant Prop. Soc.*, **28**, 539–545.

Zimmerman, R. H. (1979) The Laboratory of Micropropagation at Cesena, Italy. *Comb. Proc. Int. Plant Prop. Soc.*, **29**, 398–400.

Zimmerman, R. H. (1984), in *Handbook of Plant Cell Culture*, Vol. 2, *Crop Species* (ed. W. R. Sharp, D. A. Evans, P. V. Ammirato and Y. Yamada), MacMillan, New York, pp. 369–395.

Zimmerman, R. H. (1985) in *Tissue Culture in Forestry and Agriculture* (ed. R. R. Henke, K. W. Hughes and M. J. Constantin), Plenum Press, New York, London.

Zohary, D. and Speigel-Roy, P. (1975) Beginnings of fruit growing in the Old World. *Science*, **187**, 319–327.

Index